Chapter 6 p. 117-134
torn out p 16/6/95
etc.

ML

# CRC SERIES IN BIOCOMPATIBILITY

Series Editor-in-Chief

**David F. Williams, Ph.D., D.Sc.**
Professor
Institute of Medical and Dental Bioengineering
University of Liverpool
Liverpool, England

FUNDAMENTAL ASPECTS OF BIOCOMPATIBILITY
**Editor: David F. Williams, Ph.D., D.Sc.**

BIOCOMPATIBILITY OF ORTHOPEDIC IMPLANTS
**Editor: David F. Williams, Ph.D., D.Sc.**

BLOOD COMPATIBILITY
**Editors: David F. Williams, Ph.D., D.Sc., and Donald J. Lyman, Ph.D.**

BIOCOMPATIBILITY OF DENTAL MATERIALS
**Editors: D. C. Smith, Ph.D. and David F. Williams, Ph.D., D.Sc.**

SYSTEMIC ASPECTS OF BIOCOMPATIBILITY
**Editor: David F. Williams, Ph.D., D.Sc.**

BIOCOMPATIBILITY IN CLINICAL PRACTICE
**Editor: David F. Williams, Ph.D., D.Sc.**

BIOCOMPATIBILITY OF CLINICAL IMPLANT MATERIALS
**Editor: David F. Williams, Ph.D., D.Sc.**

BIOCOMPATIBILITY OF TISSUE ANALOGS
**Editor: David F. Williams, Ph.D., D.Sc.**

# Biocompatibility of Tissue Analogs

## Volume I

Editor

**David F. Williams, Ph.D., D.Sc.**
Professor
Institute of Medical and Dental Bioengineering
University of Liverpool
Liverpool, England

CRC Series in Biocompatibility

Series Editor-in-Chief

**David F. Williams Ph.D., D.Sc.**

CRC Press, Inc.
Boca Raton, Florida

**Library of Congress Cataloging in Publication Data**
Main entry under title:

Biocompatibility of tissue analogs.

   (CRC series in biocompatibility)
   Bibliography: p.
   Includes index.
   1. Biomedical materials.   2. Tissues.  I. Williams,
D. F. (David Franklyn)  II. Series.  [DNLM: 1. Bio-
compatible Materials.   2. Histocompatibility.
QT 34 B5585]
R857.M3B563  1985    610'.28     84-14974
ISBN 0-8493-6634-8 (v. 1)
ISBN 0-8493-6635-6 (v. 2)

Direct all inquiries to CRC Press, Inc., 2000 Corporate Blvd., N.W., Boca Raton, Florida, 33431.

© 1985 by CRC Press, Inc.
International Standard Book Number 0-8493-6634-8 (Vol. I)
International Standard Book Number 0-8493-6635-6 (Vol. II)

Library of Congress Card Number 84-14974
Printed in the United States

# PREFACE

Several years ago a CRC Uniscience Series on Biocompatibility was conceived and during the years 1981 to 1983, numerous volumes have been published in this series. These covered some of the basic principles of biocompatibility under the headings *Fundamental Aspects of Biocompatibility* (Volumes I and II), *Systemic Aspects of Biocompatibility* (Volumes I and II), and *Biocompatibility of Clinical Implant Materials* (Volumes I and II), and also focused on clinical areas with *Biocompatibility of Dental Materials* (Volumes I to IV), *Biocompatibility of Orthopedic Implants* (Volumes I and II), *Biocompatibility in Clinical Practice* (Volumes I and II), and a future volume on *Blood Compatibility*.

With the increasing acceptance of the importance of biocompatibility in implanted devices, dental materials, and medical engineering in general, further volumes in this series are now planned. In the near future a volume on *Techniques for Biocompatibility Testing* will be published and other volumes on specific themes of biocompatibility will be added as the subject progresses.

The present volume deals with one very topical aspect of biomaterials that has received minimal coverage in the preceding volumes of the series. This is the use of natural tissues and their analogs for tissue reconstruction. It is widely recognized that tissues themselves provide the best type of material for reconstructive surgery; indeed, viable tissues in the form of transplants can be extremely effective. There are, however, many technical, biological, legal, and ethical constraints with tissue transplantation and there has been a strong emphasis in biomaterials research in just the last few years on the preparation of treated natural tissue and synthetic analogs of these tissues for reconstructive procedures. This volume aims to cover the aspects of biocompatibility that relate to the use of these materials for the augmentation, repair, or replacement of both soft and hard tissue.

As with other parts of the series, the rationale and planning are discussed in the first introductory chapter and there is little need to preempt that discussion in a lengthy preface. It is hoped that these volumes will complement those others that have dealt with the more traditional biomaterials in this series and thus assist in the continuing development of this subject of biocompatibility. These volumes could not have been completed without the expertise of either the contributing authors or the editorial staff at CRC Press. To all these I wish to express my gratitude.

**David F. Williams**

# THE EDITOR

**David F. Williams, Ph.D., D.Sc.,** is Professor and Director of the Institute of Medical and Dental Bioengineering at the University of Liverpool, England.

Dr. Williams received a B.Sc. with 1st class Honors in 1965 and a Ph.D. in 1969, both in Physical Metallurgy at the University of Birmingham. He has been on the staff at Liverpool University since 1968, apart from a year as Visiting Associate Professor in Bioengineering at Clemson University, South Carolina in 1975. He was awarded the degree of D.Sc. in 1983.

Dr. Williams is a Chartered Engineer and a Fellow of the Institution of Metallurgists. He has been a council member of the Biological Engineering Society. He was presented with the Clemson Award for contributions to the literature in biomaterials by the Society for Biomaterials in 1983. He is currently editor of *CRC Critical Reviews in Biocompatibility* and is European editor for the *Journal of Biomedical Materials Research*.

Dr. Williams has major research interests in the interactions between metals and tissues, polymer degradation, and the clinical performance of biomaterials. He has authored many research papers, reviews, and books in these areas.

# CONTRIBUTORS

**James M. Anderson**
Professor of Pathology, Macromolecular
  Science, and Biomedical Engineering
Departments of Pathology and
  Macromolecular Science
Case Western Reserve University
Cleveland, Ohio

**Praphulla Kumar Bajpai**
Professor
Department of Biology
University of Dayton
Dayton, Ohio

**H. P. Dinges**
Doctor
Ludwig Boltzmann Institute for
  Experimental Traumatology
Vienna, Austria

**Anne Hiltner**
Professor
Department of Macromolecular Science
Case Western Reserve University
Cleveland, Ohio

**R. F. Oliver**
Senior Lecturer
Department of Biological Sciences
Dundee University
Dundee, United Kingdom

**Kari U. Prasad**
Research Associate Professor
Laboratory of Molecular Biophysics
University of Alabama at Birmingham
Birmingham, Alabama

**H. Redl**
Docent Doctor, Diploma in Engineering
Ludwig Boltzmann Institute for
  Experimental Traumatology
Vienna, Austria

**Eric E. Sableman**
Consulting Biomedical Engineer
Rehabilitation R & D Center
Palo Alto VA Medical Center
Palo Alto, California
            and
Bio-Response, Inc.
Hayward, Calif.

**G. Schlag**
Professor and Doctor
Ludwig Boltzmann Institute for
  Experimental Traumatology
Vienna, Austria

**Karen L. Spilizewski**
Research Assistant
Department of Macromolecular Science
Case Western Reserve University
Cleveland, Ohio

**Dan W. Urry**
Director
Laboratory of Molecular Biophysics
University of Alabama at Birmingham
Birmingham, Alabama

**David F. Williams**
Professor
Institute of Medical and
  Dental Bioengineering
University of Liverpool
Liverpool, England

# TABLE OF CONTENTS

Chapter 1

# NATURAL TISSUES AND THEIR SYNTHETIC ANALOGS AS BIOMATERIALS: AN INTRODUCTION

**D. F. Williams**

## I. INTRODUCTION

The tissues of the body are subjected to numerous diseases and injuries which, if untreated, may lead to pain, loss of function, restricted mobility, disfigurement, and severe disability. In many cases, the preferred course of treatment involves removal of the affected tissues and their replacement by some substitute. In considering the nature of the substitute to use, the surgeon is presented with a number of alternatives. In principle, the first decision involves the choice between a viable tissue graft or a nonvital component, i.e., whether to use a transplant or an implant. In practice, this is rarely a serious question as the indication for, and availability of transplants and implants are, at this stage, complementary rather than competitive. Of the more common types of viable tissue grafts, both skin and bone grafts have until recently been unchallenged as reconstructive materials since synthetic, nonvital alternatives have not been available. Similarly, with major organ transplants such as the kidney, liver, and heart, no implantable artificial organs have become functionally competitive with the natural organ replacement. In arterial replacement, the indications for viable grafts are usually a little different than those for prostheses and even with corneal replacement, corneal transplants and keratroprostheses are sufficiently different for there to be no real dilemma facing the surgeon.

The situation is changing, however, and as biomaterials science becomes more sophisticated, the possibilities of developing materials and devices to rival the performance of grafted tissues and organs become greater. It is therefore relevant to consider the relationship between transplanted tissue and implanted biomaterials and, in particular, to consider how each approach to tissue reconstruction can benefit from the lessons learned with the other.

There can be no doubt that at the present time a transplanted tissue or organ, when available, is preferable to an implanted synthetic material or device. There are limitations, however, in respect to transplanted or grafted tissues, and discussions of rejection phenomena and organ availability are so commonplace that they need take no space here. While tremendous progress has been made with immunosuppression so that tissue rejection is becoming less problematic and increasing public awareness implies a more favorable situation with donor organs, the logistical limitations still mean that transplanted vital tissues from one human to another can play only a relatively minor role in the overall surgical reconstruction program.

There are also problems of availability when homografts are involved for there is a very restricted amount of tissue available for grafting from one part of the body to another. This can easily be seen with bone grafts where only small quantities of material can be obtained. Moreover, serious questions on the risks of morbidity involved in obtaining rib or iliac crest bone arise and these have to be balanced with the benefits that are likely to accrue.

On the basis that implanted materials and devices are required for reconstructive surgery, the question remains as to which are the most suitable materials. This, of course, is the central theme of biomaterials research which, over several decades has yielded an impressive array of synthetic engineering materials for the armamentarium of the reconstructive surgeon. Many of these materials and their clinical uses have been discussed in earlier parts of this Biocompatibility Series.

The use of these synthetic materials and devices itself has its limitations. It has largely been based on the assumption that an irreversible destructive process has taken place and that the best solution is to find an inert material to physically replace and functionally imitate as far as possible the affected tissue. The replacement of natural living tissue by inert, nonbiological materials clearly has serious shortcomings and even the most clinically successful of synthetic prostheses must, by definition, be structurally and functionally naive. As we look to the more complex prostheses and reconstructive procedures, so this deficiency will become more apparent.

If the use of vital natural tissues from the patient himself or from the donors is logistically and immunologically difficult and if the use of synthetic nonbiological material is severely restricted in the type of tissue and function that they can replace, what is the logical step to produce more readily available, biologically acceptable, and structurally compatible materials? The answer to this question comprises the theme of this volume for it concerns natural, but not vital, tissues and their synthetic analogs. We shall, in fact see how there has been a trend to the development of such materials using two principles that are different in concept but similar in result.

All tissues are complex structures containing a multitude of cellular and extracellular components. Differences between components are often quite subtle but of great significance in determining their performance. Given the availability and immunological problems of tissue grafts, a possible approach to the situation is to consider natural tissues taken from donors, where supply is no problem (i.e., using tissues of animal origin) and where antigenicity, sterility, and storage problems can be obviated by chemical or other pretreatments. Thus, animal tissues pretreated by agents such as glutaraldehyde and often freeze-dried, have emerged as potential biomaterials. Clearly, the materials will be nonvital and therefore cannot simulate exactly the natural tissues, and the pretreatment and storage may affect the complex interactions between components referred to above so that mechanical, physical, and chemical properties change, but they do offer the potential of being structurally analogous and functionally non-to-dissimilar to the tissue that is being replaced.

This problem has also been approached from the opposite direction in which detailed consideration is given to these structural components of natural tissue and attempts are made to reproduce them by totally synthetic routes. It is still impossible to conceive rebuilding even the extracellular components of tissues exactly from synthetic materials to give all of the structural characteristics, but it is possible to select some of these components, perhaps the simplest and hopefully the most important, to produce a realistic tissue-like synthetic material. Thus, polyaminoacids and polypeptides have been prepared for implantation in soft tissue reconstruction. Similarly, the mineral hydroxylapatite has been prepared to approximate to the mineral phase of bone for hard tissue reconstruction. It is quite possible, of course, for these two different approaches to lead to the same type of product and they must be considered as two alternatives with a common objective. With the pretreatment of donor bone tissue (e.g., a xenograft obtained from oxen or calves) it is possible to remove either the mineral or organic phase. Removal of the organic phase is more common, as this obviates the antigenicity, leaving a material consisting of the mineral phase. This is chemically very similar to the most favored 'synthetic bone' product that is currently in clinical use and which, as mentioned above, consists of pure calcium hydroxylapatite, the major phase of bone mineral.

The discussion so far has assumed that the material used for reconstruction should always be stable and that the reconstructed tissue should always be identified as such. This need not necessarily be so, and we have to introduce the concept of graft incorporation, reorganization, and remodeling. For major organ replacement, we would not expect a transplant to rapidly change, structurally, as it becomes incorporated into the recipient, but would expect it to remain identifiable as the donor organ. On the other hand, simpler tissues such as skin or bone, either autogenous or otherwise, will slowly be recognized, becoming truly incorporated into the new tissue site. In some cases we can, in fact, consider the graft as a matrix for new tissue regeneration, the original graft being slowly replaced by this remodeling tissue.

With this in mind, we have to consider whether we want our replacement tissue analog to act as a permanent replacement or as a resorbable scaffold that assists in the reconstruction process of the body. There are some merits in the latter concept as this could lead to true reconstruction with the ideal replacement material. Under some circumstances, however, this can be a disadvantage, and we have to bear in mind the reasons for the tissue loss. If this loss is due to disease or an abnormal physiological process which is difficult or impossible to eradicate or ameliorate, then the replacement of the host tissue with an identical tissue may only provide a palliative, the reconstructed tissue being subjected to the same destructive conditions. A reconstructed periodontium may rapidly reresorb if peridontal disease is still present, and an augmented alveolar ridge will continue to resorb if new bone is used for the reconstruction. Under these circumstances, a replacement which is not identical to the natural tissue but is of greater stability may be preferable and this is one reason why synthetic apatites may be superior as bone reconstructing materials in comparison with new bone itself. A series of conflicting arguments are therefore developing and it is not at all exactly clear in which direction we should be going; indeed, different clinical problems may require different directions of approach. It is hoped that this volume will provide background data and reviews of experimental and clinical experiences to provide an assessment of the state-of-the-art in natural tissues and their synthetic analogs for reconstructive surgery.

In this volume are examples of treated natural tissues, totally synthetic materials, and hard and soft tissue reconstruction. One basic problem underlying many of the applications of treated natural tissues is that of the immunological sequences of implanting tissue derived from human donors or other species; Chapter 2 discusses the background and present knowledge concerning this very important subject. The following five chapters deal with various aspects of preparation, properties, biocompatibility, and clinical uses of natural organic materials, including collagen, polyamino acids, the polypentapeptide of elastin, and fibrin. Bone induction by implanted material is discussed in the first three chapters of Volume II, with contributions on grafts, demineralized bone matrix, and calcium phosphate ceramics. Combinations of different types of material to produce composite tissue analogs are the subject of the following two chapters, with contributions on novel collagen-plastics composites and on hydroxylapatite-polymer composites as bone replacement materials.

The final five chapters all refer to the use of natural or pseudonatural materials in the cardiovascular system, with discussions on natural tissue arterial prostheses, phospholipid polymers for membranes and other blood contacting surfaces, polysaccharide resins, and bioprosthetic heart valves.

Chapter 2

# IMMUNOLOGICAL ASPECTS OF TREATED NATURAL TISSUE PROSTHESES

**Praphulla Kumar Bajpai**

## TABLE OF CONTENTS

# I. INTRODUCTION

The term "bioprosthesis" means an artificial part produced by the joint efforts of nature and man.[1] Thus, free grafts of autogenous tissue as well as physically and chemically preserved natural tissues can be classified as bioprostheses. Currently, bioprostheses are being used as scaffolds, patches, conduits, valves, and in some cases, entire organs to replace defective parts of the human body.

Some of these bioprostheses are alive while others are dead at the time of implantation. The host uses different effectors of the immune system to remove these devices from its confines. For example, a physically or chemically treated dead tissue is removed by enzymatic degradation and/or phagocytosis rather than by lymphocyte-mediated cell cytotoxicity. Since the text of this chapter deals with the immunogenicity of treated natural tissue prostheses, it seems appropriate to outline the current concepts of how the immune system functions in higher vertebrates and humans.

# II. IMMUNE SYSTEM

The defense system of the body in most instances resists the acceptance of any organic matter (foreign substance) not synthesized by its deoxyribonucleic acid (DNA) via the immune system.[2] Two kinds of effector mechanisms mediate immune responses.[3] Some immune responses are mediated by circulating specific globular proteins called "antibodies". Antibodies are produced by differentiated B lymphocytes known as plasma cells. The B cell can accomplish this task alone in the case of certain polymeric forms of antigen insults (thymus-independent responses) or with the help of another type of lymphocyte known as the helper T cell (thymus-dependent responses). These types of humoral immune responses are directed toward a wide range of molecular and cellular foreign substances.[2-4]

The second way in which the immune system deals with foreign material (antigens) is by means of leukocytes. All the leukocytes of the blood participate in cell-mediated immunity (CMI).[3] However, the specificity of CMI response depends on a special class of lymphocytes called the "T lymphocyte" or "T cell". There are two functionally different forms of CMI. One is known as cell-mediated cytotoxicity and the other as delayed-type hypersensitivity. These are mediated by two different subpopulations of T lymphocytes. Cells classified as $T_c$ or $T_k$ (cytotoxic) cells have surface markers Leu—2a, 2b or OKT-5,8 in humans.[2] These cells recognize and react specifically with antigens located on the surface of live target cells and kill them by contact and/or lysis.[5] The exact mechanism by which this is accomplished is still not known.

Delayed-type hypersensitivity (DTH) is mediated by a small population of $T_D$ cells and a large population of nonsensitized effector cells. Human $T_D$ cells have surface markers Leu-1 or OKT-4.[2] The sensitized $T_D$ cells, on reacting with conventional antigens such as polymerized bovine serum albumin or cell surface (live or dead) antigens, secrete lymphokines. The lymphokines then recruit a variety of accessory cells and especially macrophages. The recruited cells (polymorphonuclear leukocytes, eosinophils, basophils, monocytes, lymphocytes, and macrophages) have no specificity for the antigen and act as nonspecific effectors of tissue inflammation.[2-6] The recruited macrophages become more reactive and phagocytic to a wide variety of antigens, cell types, and inert particles.[2] Most immune responses usually involve the activity and interplay of both humoral and CMI responses.[2-4]

A large majority of tissue bioprostheses at the time of implantation are lifeless (do not have flowing cell membranes). Hence, the likelihood of inducing a cytotoxic type of CMI against these bioprostheses is nonexistent and tests designed for testing cytotoxic responses of CMI will not, in these cases, yield meaningful information. On the other hand, chances of inducing humoral and/or DTH-type of CMI depend on various factors, including conformation, quantity, and site of antigen introduction. The final step in antigen, lymphocyte, and macrophage interaction is expressed by the development of DTH reaction.[4] For manifestation of DTH at the site of implantation, good lymphatic drainage as well as vascularity are essential, since the initial contact between the antigen and $T_D$ cells is made in the draining lymph nodes. Since degradation products of several bioprostheses can provide antigen to the entire lymphatic system, skin tests usually provide a good indication of induction of CMI. Skin has excellent lymphatic drainage and vascularity. However, noncutaneous and systemic DTH reactions also occur in tissues other than skin. In individuals highly sensitized to tuberculin, injections of tuberculin can cause severe inflammation and necrosis of cornea as well as systemic reactions characterized by headache, malaise, and prostration. DTH is rarely fatal in humans.[5]

## III. PRESERVATION AND TREATMENT OF NATURAL TISSUES

Microbial free transition of live or dead tissue prosthesis from the source to the host is essential for successful functioning of the bioprosthesis. Live donor homografts or allografts are usually treated with antibiotics and/or glycerol and stored during transit at subzero temperatures. Gas sterilization, radiation, and ultraviolet irradiation have also been used to preserve both xenografts and allografts. Xenografts are usually preserved by treating them with formaldehyde and/or glutaraldehyde.

Since formaldehyde has only one aldehyde group available for cross-linking proteins,[7] the binding of formalin to protein is unstable and often leads to reversal of collagen cross-linking in vivo.[8] Glutaraldehyde, on the other hand, has proven to be a highly effective protein cross-linking agent.[9] The two aldehyde groups of glutaraldehyde give it the capability of reacting with various side chains of amino acids.[10,11]

The ability of glutaraldehyde to stabilize tissue proteins has led investigators to assume that glutaraldehyde treatment renders biological material inert, nonviable, nonantigenic, and nonimmunogenic.[15-17] However, glutaraldehyde has been used to stabilize antigenic determinants, prepare immobilized but active enzymes, vaccines, polymeric immunogenic allergens, and induce humoral and/or CMI responses.[13-15,18-20]

Some of the factors which favor the induction of an immune response are carbohydrate, protein, lipoprotein, glycoprotein, or lipopolysaccharide conformation, accessibility to determinants, large size and surface area, digestibility, higher stability, and polymeric conformation.[13-15] Glutaraldehyde cross-links and stabilizes proteins,[21-25] polymerizes biologic material,[26-28] masks some and unmasks other determinants,[13-15,29,30] and delays,[12,31,32] but does not prevent the digestion process.[13-15,33,34] Hence it is logical to expect the glutaraldehyde-treated tissue valves should evoke an immune response. Formaldehyde-treated antigens have been routinely used to produce effective vaccines.

## IV. IMMUNOGENICITY OF BIOPROSTHESES

Usually, generation of an immune response against a bioprostheses or its contents are regarded as undesirable. However, antibodies as well as antigen-antibody complexes can protect grafts from the cellular attack mounted by the host.[2] On the other hand, in certain instances it may be desirable to resorb the implant and replace it with the tissue of the host. In the latter case, assistance by the immune system can hasten the resorption of the implant. Both humoral and cellular (DTH) responses are induced in amoebiasis, but neither the quality nor the magnitude of immune responses can be correlated with the state of the disease caused by *Entamoeba histolytica*.

Since the induction of immune response and its role in the survival of the bioprosthesis and/or the host is often controversial, a set of questions should be asked by the biomaterial scientists and/or surgeons in relation to the immunogenicity of the desired bioprosthesis:

1.    Is the bioprosthesis or its contents immunogenic?
2.    If the device or its contents are immunogenic, can they break the tolerance (nonreactivity) to self-antigens by inducing an immune response?
3.    Is the device degradable by the cellular system of the host? If so, what is the time course of degradation?
4.    If the device or its contents are immunogenic and slowly degradable, can the small or trace amounts of antigen released induce inflammatory reactions away from the implant even if the immune response cannot effectively degrade the bioprosthesis?
5.    If the device or its contents are immunogenic, can the immune system increase its rate of degradation or protect it from cellular attack?
6.    Is the site of implant accessible to the cells of the immune system and/or the nonspecific cells recruited by the immune system?
7.    What is the time course of one or more of the above events in relation to the projected life of the device and/or the host?
8.    Will the reintroduction of a similar device enhance or suppress one or more of the above events?

## A. Bone

An ideal bone autograft should be biocompatible, nonimmunogenic, and replaceable by the host. The best material to correct discontinuity defects in bone is a bone autograft.[35-39] Massive replacements of bone cannot be easily achieved by bone autografts.[40] Involvement of additional surgery, higher rate of resorption, and lack of adequate vascularization in some instances have also created serious problems with the use of bone autografts.[41-43]

### 1. Allograft (Homografts)

Availability of allograft bone in various sizes and shapes makes it easier to match the size and shape of the host bone.[44] However, allografts are likely to trigger the immune system of the host. To circumvent or minimize the immune response, bones have been chemically treated, boiled, irradiated, frozen, freeze-dried, decalcified, and demineralized.[39-45]

Freezing and freeze-drying of allografts results in the loss of allograft immunogenicity.[46,47] According to Cloward,[44] the percentage of "take" in the interbody fusion with gas-sterilized cadaver bone equaled or exceeded that of bone autografts. After 90 days of implantation, complete resorption and replacement of formaldehyde-treated allograft bone with development of new osteoid tissue was reported by Goncalvez and Merzel.[48] Freeze-dried bone allografts of fine particle size have also been used to treat periodontal defects.[49]

Both frozen and fresh allogeneic cortical and cancellous bone have been observed to induce cellular and humoral responses.[50-52] Burwell[53] reported that freeze-dried bone allografts after 3 to 6 weeks of implantation are quite dense and from then on undergo very slow resorption and replacement with new bone.[39] Vascularization begins more slowly in preserved bone allografts than in autogenous bone because of host vs. graft response and density of mineralized cortical bone.[54] Since the allogeneic bank bone, when deep-frozen or lyophilized, is still antigenically active[51,52] it cannot be considered as an ideal substitute for autologous bone graft.[39] However, blocking antibodies could play a role in protecting the bone allograft.[39,50]

### 2. Xenografts (Heterografts)

Bovine Kiel (keel) bone has been successfully used in many European hospitals without immunological complications.[55-59] Taheri and Gueramy[60] reported equal degrees of success with autogenous and defatted-sterilized calf kiel bone. However, Goncelvez and Merzel[48] reported that formaldehyde-fixed bone heterografts, on implantation in rats, behaved as foreign bodies. In general, xenograft studies have not been fruitful.[39]

## B. Cartilage

Nonvascular and dense matrix of the articular cartilage virtually eliminates its ability to induce immune responses.[40,61-64] The chondrocytes of the cartilage are embedded in a non-vascularized matrix and receive their nutrition by diffusion[64] and are not easily accessible to the effectors of the immune system.

Low antigenicity and an adequate supply of preserved allogeneic cartilage favor its use for augmenting bony and cartilagenous structures such as dorsum, nose, chin, facial bones, external ear, and ossicular bones. However, allogeneic cartilage is antigenic and is rejected by the effectors of CMI.[64] Whether the antigenic activity is in the chondrocytes or in the matrix is still a matter of controversy.[64] However, graft survival seems to depend on the cartilage having live chondrocytes included in an intact matrix. Cartilage grafts with dead chondrocytes are rapidly resorbed on implantation.[64]

Experiments conducted by Smith[65] suggest that on implantation, live chondrocytes surround themselves with a matrix so that the host does not have time to mount a cellular attack.[64,65] However, there is always the remote possibility that chondrocytes are nonantigenic.[64]

According to Meijer and Williams,[64] if the chondrocytes are preserved (alive) there is no difference between the survival of allogeneic and autogenous cartilage. Some investigators have even used bovine cartilage with initially good results.[64,66,67] However, according to deCarvalho and Okamoto,[68,69] rib cartilage autografts are better accepted than rib cartilage allografts. They also reported that the host eliminated the allograft cartilage at a higher rate than the autogenous cartilage and that the rate of healing was retarded in bone reconstructed with allograft cartilage.[68,69] In contrast, a follow-up study on implants of irradiated lyophilized allogeneic cartilage suggests that only 10 to 20% of resorption occurred over a duration of 1 to 5 years.[64] Peer[70] reported that it took a minimum of 2 years to absorb stingray cartilage implanted s.c. in the abdominal tissues of humans.

## C. Decalcified and/or Demineralized Bone

The success of a clinical bone graft depends on its ability to stimulate formation of bone. The chances of survival of bone cells on their removal from the original blood supply is quite remote.[45] Hence, the cell stimulus has to be provided by the host.[39,41,45,53] Decalcified bovine cortical chips were used as early as 1889 to fill osteomyletic cavities.[54,71] Since then, both xenogeneic and allogeneic preparations obtained from decalcified and/or demineralized bones have been used to induce bone growth.[71-80]

According to Urist et al.[81] digestion and extraction procedures destroy the highly antigenic cellular membranes and soluble haptenic glycoproteins while the remaining collagen remains as a weak antigen. Collagen plays a major role in the mineralization of bone. Usually, collagen implants ossify within 4 to 6 weeks.[82] The antigenicity of the decalcified bone does not play a critical role in the survival of the graft since the graft itself is eliminated and the site of implantation is remodeled with new bone.[39] However, what effects the generation of an immune response will have against organic constituents of other connective tissue organs is a matter of conjecture and concern. The risk of inducing an autoimmune response in an otherwise healthy individual should not be overlooked, especially when promising nonorganic and nonimmunogenic substitutes are being introduced for rebuilding bone.

## D. Collagen

Purified and/or cross-linked (stabilized) collagen preparations have been used as hemostatic agents and biological dressings. They have also been used in management of burn wounds, ophthalmology, orthopedics, and oral, dental, hand, and plastic surgeries.[25]

Collagen Type I and Type III are found in the skin and parenchyma of several organs. Collagen Type II is a component of cartilage. A hydroxyl-rich Type I collagen has been isolated from porcine heart valves,[83] and 50% of the dry weight of bovine heart valves is composed of collagen.[84] Collagen is also a major component of vascular tissues.[13,14]

Each tropocollagen (monomer) molecule has short nonhelical amine and carboxy terminal amino acid sequences apart from the rigid helical structure.[25] These nonhelical portions, called "telopeptides", are cross-linked by intra- and intermolecular bonds.[25] The branched telopeptides of the collagens contain several primary immunogens.[25,85,86] Crossreactivity and the ability of denatured collagen to induce cellular sensitivity indicates that helical antigenic determinants are not involved in eliciting the above responses.[87,88] Collagen is a poor antigen and/or immunogen.[24,25,89,90] However, on implantation, collagen induces a chronic inflammatory reponse and eventually is degraded by the host.[91] Homologous, heterologous, or denatured Type I collagens are capable of inducing both humoral and cellular immune responses.[87] However, Type I and III collagens are less immunogenic than Type II collagens.[88] Immune responses against collagens has been demonstrated by a variety of serological procedures, including passive hemagglutination, complement fixation, immunofluorescence, radioimmunoassay, blastogenesis, skin reaction, and macrophage migration inhibition.[88,92]

Stenzel et al.[93] reported that pure collagen implants do not elicit a severe immune response.[24] However, Zyderm-Cl has induced unusual responses, both adverse and positive, in a very small percentage of highly screened participants.[24]

Low antigenicity and poor inflammatory response of collagen prompted several investigators to use collagen devices as substrates for tissue ingrowth for repairing bone defects.[71-76,78-82,94-96] However, significant impairment of wound healing[94] and pronounced inflammation and retardation of bone formation has been reported by other investigators.[98]

Enzymatic digestion of collagen results in a substantial loss of antigenic determinants of tropocollagen.[25,85,99,100] Enzymatic digestion also reduces the tyrosine and aromatic amino acid antigenic components of nonhelical peptides in collagen.[25,95] Zyderm, a collagen preparation which is digested, purified, and sterilized by pepsin, has been shown to be a poor inducer of both humoral and CMI responses.[24] According to several investigators, cross-linking of collagen by treating it with radiation, UV irradition, and/or glutaraldehyde reduces its immunogenicity.[24,25,85,99,100] The data collected in our laboratory suggests that glutaraldehyde treatment of tissues containing high amounts of collagen does not reduce the immunogenicity of the treated tissue but alters the specificity of the antigenic determinants. These altered determinants induce specific immune responses, and when tested against untreated native or original tissue antigens, showed reduced reactivity.[13-15]

Isologous and homologous muscle fragments treated with 2% glutaraldehyde and sealed in polyethylene tubes induced, on implantation in rats, a chronic inflammatory immune-type response characterized by the presence of eosinophils in association with lymphocytes.[101,102] Immunizations with glutaraldehyde-polymerized bovine serum albumin results in the formation of two types of antibodies, one specific for the protein, the other specific for the glutaraldehyde-protein complex.[103]

## E. Fibrin

Compressed and molded ox fibrin can be used clinically to correct tissue weakness and deficiency.[104] Compressed and molded fibrin is resorbed by polymorphonuclear digestion within 4 to 8 weeks of implantation.[104] According to Horvath et al.,[105] the antigenicity of ox fibrin is quite poor. Caperauld[104] reported that repeated exposure of some rats to compressed and molded ox fibrin at intervals of 4 weeks does not induce allergic reaction.

## F. Skin

Skin is the best substitute for skin allograft.[106] However, these allografts must be removed from the skin before immune rejection occurs.[106] Tanned human skin allografts are also rejected. However, treatment of human skin allografts with 0.25% glutaraldehyde does prolong the retention time from 10.9 to 21.8 days.[32] Frozen, irradiated porcine skin xenografts have been claimed to be equally effective.[107] Several collagen preparations are now being used as skin dressings. Immunogenic potentials of these were discussed earlier in the section on collagens. Mammalian skin collagens show extensive crossreactions.[13,14]

## G. Conduits

Major components of vascular tissues are collagen, elastin, mucopolysaccharides, glycoproteins, and soluble proteins. Collagen forms the bulk of the tissue.[13,14]

The ideal vasular graft is a healthy venous autograft. Before using stabilized vascular allografts or xenografts, several investigations were conducted with allografts.[108,109] Usually, the procedure used to investigate the immunogenicity of these grafts dictated the conclusion whether the allograft was immunogenic or mildly antigenic. However, the general consensus is that vascular allografts are immunogenic.[15]

### 1. Veins

Perloff et al.[110] reported that venous allografts were immunogenic, since they induced cellular immune responses similar to those produced by skin allografts and other solid organs. Venous allografts are mildly antigenic.[111] Schwartz et al.[111] observed that grafting of allograft vessels did not accelerate the rejection of subsequent skin grafts and did not decrease the titers of complement.

The major concern in the success of a venous graft is maintenance of patency. Induction of immune response does not necessarily mean that the vessel will be occluded.[110] However, rejection does seem to play a role in the failure of long-term patency. The damage caused by the immune system eventually exceeds the rate of proliferation of donor epithelium.[112]

Dogs have natural antibodies against human erythrocytes and thus are often used to test the success of human tissues.[113] Esato et al.[114] used human umbilical cord vein as canine arterial by-pass grafts and reported that the immunogenicity of the grafts was effectively suppressed by cross-linking the veins with 0.5 to 1% glutaraldehyde.[14,15] Shiel et al.[113] used tanned human saphenous veins for reconstructing segments of canine arteries. They reported that the amount of immunological involvement was minimal, but the graft was gradually replaced by the connective tissue of the host.

Antigenic challenge of sensitized lymphocytes can induce the generation of biologic mediators known as lymphokines.[2-4] Lymphocyte-fibroblast activating factor (FAF) helps in the synthesis of collagen, and a lymphokine-macrophage activating factor (MAF) helps in the destruction of collagen.[115] Since stabilization of the tissue by glutaraldehyde delays its rate of degradation, it is possible that FAF-induced formation of new connective tissue overcomes the rate of degradation.[14,15] Induction of immune reponse as well as encapsulation of glutaraldehyde-stabilized porcine aortic valve and bovine pericardial tissue xenografts in rabbits has been observed in my laboratory.[12-15]

### 2. Arteries

In the early 1960s, arteries obtained from cattle were used to manufacture arteriografts. Animal studies conducted with bovine arteriograft showed a high incidence of aneurysms.[109] Demonstration of lack of immune response against supernatants (containing 6.4 mg% protein), obtained from insolubilized tanned bovine artery homogenate,[116] provided the incentive for the use of tanned bovine arteriografts in the late 1960s and early 1970s.[14,15] Mattila and Fogharty[117] also reported reduced antibody response against bovine arteries treated with 1.3% dialdehyde starch and porcine arteries treated with 0.2% glutaraldehyde. An important fact, that untreated bovine artery extract supernatants had a protein concentration of 364 mg% and that serological tests were conducted against untreated extracts by Defalco,[116] was overlooked by other investigators,[117-119] who followed similar procedures to test for immunogenicity against tanned tissues.[12-15] Early incidence of thrombosis, high rate of infection, aneurysmal degeneration, lack of length, inadequate diameter, and high cost has discouraged the use of tanned bovine arteriografts.[120]

Okamura et al.[121] tanned ascending aortas with valves obtained from racoon-dogs and cats. On implanting these tanned tissues in the descending aortas of dogs, they observed that primary and secondary tanned aorta grafts induced a lower titer of antibodies that untanned grafts of aorta. Hemagglutination and hemolysis titers were measured using racoon-dog and cat erythrocytes. Okamura et al.[121] did not test the sera obtained from dogs implanted with tanned aortas against extracts of tanned aortas.

Glutaraldehyde cross-links and insolubilizes biological material.[23] Hence, the supernatants of such extracts can only contain trace amounts of insolubilized antigen. Tests, including skin reactions, conducted with such supernatants, at best can reveal very little information about the immunogenicity of tanned tissues.[15] Using fine particulate antigen preparations of tanned bovine pericardial tissue, tanned porcine aortas, and tanned porcine aortic valves, we have shown induction of humoral as well as cellular immune responses against tanned tissues in rabbits.[12-15,122]

## H. Heart Valves

Major components of the fibrous network of heart valves are mucopolysaccharides, structural glycoproteins, soluble proteins, elastin, collagen, and cells.[8] The mucopolysaccharides contain 65% hyaluronic acid, 25% chondroitin sulfate A/C, and 10% chondroitin sulfate B.[13,14,123] In 1960, Harker and Starr[17] successfully replaced mitral and aortic valves by means of prosthesis. The ideal valve substitute has yet to be devised. More than 50 totally mechanical or tissue prosthetic heart valves have been introduced during the past 2 decades.[17]

### 1. Allografts

The first operation to replace a defective heart valve with a fresh aortic allograft valve was performed in August 1962.[17,124] Since then, preserved aortic allografts and allograft valves fabricated from other tissues, such as fascia lata, dura mater, and pericardium, have also been evaluated. Allografts are immunogenic. However, since the ingrowth of host tissue into the allograft replaces the leaflet with new viable and pliable leaflet, the role of immune response is of academic importance. Allograft valves are still used fairly extensively in the U.K. and New Zealand.[17,124] Difficulty in procuring the bioprosthesis rather than poor results is responsible for the infrequent use of allografts today.[125] Thiede et al.[126] studied the immunogenicity of allogeneic valve leaflet transplants in syngeneic, weakly allogeneic, and strongly allogeneic rats. Leaflets were transplanted in a free-floating manner in the abdominal aorta and antibodies were measured by hemagglutination procedure, while CMI response was determined by observing rejection of leaflet donor-specific skin. Weakly allogeneic systems showed induction of humoral response by a sensitive hemagglutination procedure and accelerated rejection of skin grafts.[126] Strongly allogeneic systems showed humoral response measured by a less sensitive hemagglutination procedure and rejection of skin grafts without development of vascularization (white graft rejection).[126] Transplants of leaflets in syngeneic systems did not induce immune responses.[126]

### 2. Xenografts

Problems with procurement and sterilization and technical difficulties in inserting the unmounted allograft valve in the subcoronary positions prompted investigators to look for xenograft substitutes.[17]

### a. Porcine Heart Valves

Carpentier et al.[8] were the first to replace human heart valves with porcine xenograft heart valves.[15-17] Soluble protein, mucopolysaccharides, structural glycoproteins, and cellular components of the porcine heart valves have the highest degree of antigenicity. Elastin and collagen components of the porcine heart valve xenograft are the least antigenic.[8]

Initial attempts to replace defective human heart valves were made with porcine heart valves preserved in organic mercurial salt solution or 4% formalin.[8,15,127] On implantation in patients, these xenografts showed thinning of the leaflets, denaturation of elastin and collagen, and inflammatory and immunologic type of cellular reaction.[8,15,127,128] However, presence of antibodies either to valve or muscle tissue was not detected in the sera of these patients by tanned cell hemagglutination and immunofluorescent procedures.[8,16] Porcine heart valves also have varying amounts of myocardial tissue supporting the right cusps.[129]

In an attempt to eliminate cells and soluble proteins and denature glycoproteins and stabilize collagens, Carpentier et al.[8] developed a procedure which included oxidation and glutaraldehyde treatment of the porcine heart valve xenografts. After washing or electrodialysis and oxidation, the xenografts were tanned with glutaraldehyde.[8,15] Carpentier et al.[8] reported that tanning of the tissue reduced the antigenicity of the porcine heart valve xenograft. However, implantation of the porcine xenograft bioprosthesis in patients resulted in a high incidence of late postoperative failure associated with perforations, rips, and calcifications.[16] This prompted Carpentier[16] to drop the metaperiodate oxidation step from the tanning procedure.[8,15,16] Major manufacturers of porcine heart valve bioprostheses currently use glutaraldehyde concentrations ranging from 0.2 to 0.625% to tan the xenografts.[14,15,127]

Since tissue valves are dead when inserted, they shrink, degenerate, and calcify with time. Glutaraldehyde treatment can delay this process but cannot prevent gradual and eventual degradation of the xenograft.[14,15,31,33] Thus, it is possible that the degraded products of the valve may be capable of evoking an immune response.

According to Frolova et al.,[118] water-soluble extracts of glutaraldehyde-treated porcine heart valve (GTPHV) were immunogenic but had a fewer number of antigens than untreated porcine heart valve extracts (UTPHV). Spray and Roberts[130] reported that both collagen degeneration and thrombosis were observed in 51 patients implanted with GTPHV bioprostheses. They also observed the presence of inflammatory cells and/or immunologically competent cells in more than 70% of GTPHV bioprostheses retrieved 2 to 25 months after implantation.[130,131] Spray and Roberts[130] suggested that GTPHV bioprostheses are probably neither bioinert nor nonantigenic.[130,131]

Several investigators have studied the immunogenicity of glutaraldehyde-stabilized or tanned biological materials.[34,101,102,116,132] However, by the late 1970s, only a few investigators had reported on the immunogenicity of glutaraldehyde-stabilized porcine heart valve xenografts.[16,118,130,131] Hence, we conducted several studies on the immunogenicity of tanned tissue xenografts in rabbits.[133-140]

In the initial study, New Zealand white rabbits were immunized with UTPHV or GTPHV extracts in Freund's complete adjuvant. Sera obtained from rabbits in either group tested positive for the presence of antibodies against their corresponding antigens by capillary tube aggregation test.[133,134] Both tanned and untanned porcine heart valve antigens reacted with anti-UTPHV and anti-GTPHV sera in complement fixation, hemagglutination, and double diffusion in gel precipitation tests.[134] Three precipitation bands were formed between anti-GTPHV sera and UTPHV antigen in immunoelectrophoresis. Neither anti-UTPHV or anti-GTPHV sera reacted with porcine liver or kidney extracts.[134] Immunoelectrophoresis data suggested that glutaraldehyde treatment retains at least three soluble porcine heart valve determinants.[134] Upon i.v. challenge, all immunized rabbits displayed anaphylactic responses characterized by changes in respiratory frequency and depth, height of P and R voltages, incidence of nodal beats, atrial fibrillation, premature ventricular contraction, or a combination thereof. Differences in intensity of anaphylactic response were not discernible between the UTPHV- and GTPHV-immunized rabbits.[133,134]

In the second investigation, New Zealand white rabbits were grafted s.c. in the abdominal region with strips of UTPHV or GTPHV xenografts. Complement fixation tests showed that sera obtained from rabbits of both groups reacted with both UTPHV and GTPHV antigens.[135]

Intracutaneous injections of UTPHV or GTPHV antigens gave positive skin reaction in rabbits grafted with either UTPHV or GTPHV grafts.[136] Lymphocytes obtained from either UTPHV- or GTPHV-grafted animals on incubation with GTPHV antigens significantly inhibited the migration of macrophages from agar droplets. The presence of cellular immune response was confirmed by both in vitro and in vivo tests.[136]

Morphologic examination of grafts 3 months after implantation indicated that GTPHV grafts were encapsulated to a greater extent than UTPHV grafts.[12] Untanned and tanned grafts showed similar histopathology. Chronic inflammation with focal areas of necrosis and phagocytosis of particulates from the implant by foreign body cells were observed in hematoxylin- and eosin-stained sections of the recovered grafts.[12] Infiltration of glutaraldehyde-treated porcine heart valve xenografts with macrophages, mononuclear cells, and lymphocytes, along with degeneration of bioprosthesis, suggest that the immune system does not spare the tanned tissue xenograft.[12,15]

Ferrans et al.[141] compared scanning and electron micrographs of untreated porcine heart valve tissue with GTPHV bioprosthesis implanted for 2 to 76 months in patients. Xenografts implanted for less than 2 months showed infiltration of plasma proteins, deposition of fibrin, entrapment of erthrocytes in surface crevices, and presence of macrophages, giant cells, and platelets. Examination of GTPHV bioprostheses implanted for more than 2 months showed progressive collagen degeneration, platelet aggregation, lipid accumulation, and calcium deposition. They suggested that collagen degradation was probably due to a reversal of glutaraldehyde cross-linkages and hydrolysis by collagenases of inflammatory or immunologically activated cells.[141,142] High incidence of GTPHV bioprosthesis failure and calcification has been observed in children.[143-147]

A summary of follow-up studies of patients who have received GTPHV bioprostheses shows the following pathological changes: incidences of collagen degradation leading to cusp perforations, interstitial edema, paravalvular necrosis, amorphous collagenous material over the surface of valves, fibrin deposit, encapsulation, calcification of leaflets, infiltration of mononuclear cells, lymphocytes and macrophages, platelet aggregation, early failure of the graft in children, pulmonary disease, cardiac myopathy, arrhythmias, and diseases unrelated to the bioprosthesis.[17,33,130-131,143-145,148-158] Immunologically, both infiltration of the implantation site by mononuclear cells, macrophages, and lymphocytes, and aggregation of platelets are important. In DTH, lymphokine-recruited, nonspecific white blood cells cause the damage to the implant and other tissues.[2-5] Embedded and soluble antigens, on binding with antibody, can activate complement.[2] Components of complement-1 have been postulated to activate platelets and induce localized intravascular platelet aggregation and/or coagulation.[2]

Sheikh et al.[19] reported that of 35 patients implanted with Hancock GTPHV bioprosthesis, 58% showed antibodies to porcine heart valve tissue antigens and 32% showed positive leukocyte adherence inhibition on incubation with porcine heart valve extracts. According to them, cardiac autoantibodies, resulting from cardiac surgery, crossreact with porcine heart valve antigens. Frolova et al.[156] have reported increases in antibody concentrations to myocardial tissue and porcine heart valve antigens and activation of leukocyte migration inhibition factor after surgery to replace defective heart valves.[156] These investigators also observed that the sera of some patients reacted with normal saline extracts of porcine heart valves prior to the operation, while patients implanted with GTPHV bioprosthesis increased their cell sensitivity to human myocardial antibodies.[156] Since porcine tissues contain blood group A antigen,[14,122] it is not surprising to see that the sera of some patients will react with extracts of porcine heart valves.[14,15] However, no attempt was made by the above investigators[156] to test the induction of humoral and CMI responses against the modified GTPHV antigens. It is likely that in patients implanted with xenografts, the degradation products can reinvoke the autoimmune response and CMI. The latter, in turn, can result in cardiac myopathies, arrhythmias, collagen degeneration, valve disruption, and/or calcification either by specific mechanisms or nonspecific effectors.

Recently, Rocchini et al.[143] reported that intracardiac placement of GTPHV bioprosthesis in children induced a foreign body immune response characterized by the presence of numerous histiocytes, giant cells, and plasma cells.[143] They also observed hypergammaglobulinemia, bone marrow plasmocytosis, and, in several cases, deposition of antibodies in valvular tissue. Based on these findings, they hypothesized that tissue damage induced by an immune response and cell ingrowth could cause calcification of the xenograft valves.[143]

MaGilligan et al.[157] reported that both UTPHV and GTPHV tissue sections showed intense binding with antibodies obtained from the sera of 5 of 27 patients implanted with GTPHV bioprosthesis. These five patients did not show GTPHV leaflet degeneration. They also observed that two explanted valves contained lymphocytes and cells resembling plasma cells which stained with anti-B-cell monoclonal antibodies.[157] However, they were unable to identify the presence of monocytes and T cells by antimonocyte and anti-T cell antibodies on the explanted valves.[157] Lack of accelerated degeneration of second set GTPHV bioprosthesis and absence of other types of cells on explanted leaflets probably prompted them to suggest that immunologic response was not instrumental in the degeneration of GTPHV bioprosthesis.[157] In contrast to the observation reported by MaGilligan et al.,[157] other investigators have observed presence of mononuclear cells, macrophages, and polymorphonuclear cells[158] and other inflammatory cells which characteristically carry out the nonspecific destruction induced by DTH reactions.[15] True rejection of dead GTPHVs does not take place since the cells are not alive. Lack of good lymphatic drainage at the valvular site could account for the accumulation of lymphocytes and lack of T cells. In DTH reactions, antigen sensitization of lymphocytes takes place in the draining lymph nodes and not at the site of antigen entry or site of implantation of graft.[2]

Villa et al.[159] studied the immunogenicity of GTPHVs in mice. Mice were given 2 injections of valvular homogenates (prepared by Potter apparatus) in aluminum phosphate gel at 10-day intervals. Sera obtained from these mice was tested for immune response by means of cytotoxicity tests using porcine lymphocytes as target cells.[159] They reported low to no immunogenic response in mice injected with GTPHV homogenates.[159] Several factors can contribute to the type of results obtained by these investigators. Villa and co-workers[159] collected the sera for testing from the mice 6 days after the second injection or after a total of 26 days. In our investigations we were unable to detect positive immune response for 50 days after repeated injections at weekly intervals. Tanned tissues are processed at a slower pace than untanned tissues.[21,160] The test used by Villa and co-workers was primarily designed to detect immune responses generated against histocompatibility (HC) antigens of the porcine heart valve. Besides HC antigens, the porcine heart valves contain conventional antigenic determinants.[15,16] Glutaraldehyde is capable of modifying or altering antigenic determinants of tanned tissues.[27,29,30,160,162] Hence, sera obtained from immunized or grafted animals or from patients grafted with tanned tissue should yield antibodies specific to modified antigenic determinants. The antibodies to tanned tissues, depending on the nature of the antigen may or may not crossreact with antigens of untreated tissues.

Villa et al.[159] also grafted fresh untanned leaflets and GTPHV leaflets in cutaneous pouches on the backs of inbred C57B1 and DBA strains of mice. They observed that the untanned leaflets underwent progessive resorption over the 50 day duration.[159] We also observed progressive resorption of untanned dead porcine heart valvular tissue implanted in cutaneous pouches of rabbits.[15] By the end of 90 days most of the UTPHV grafted was resorbed.[15] Villa et al.[159] reported that the tanned tissue did not resorb and that there was no infiltration of the tanned leaflet by inflammatory cells. However, the tanned leaflet was surrounded by lymphocytes and granulocytes.[159] We also did not observe thinning of the tanned graft but did observe erosion of the borders of GTPHV graft by inflammatory cells.[12,14,15]

### b. Pericardial Tissue Valves

Serosa of bovine pericardium is lined with a monolayer of flattened cells supported by loose areolar tissue. Fibrous pericardium lies below the serosa and consists of a dense network of collagen bundles interspersed with blood vessels and elastin fibers. Pericardial tissue also contains connective tissue cells consisting mainly of elongated fibroblasts and some histiocytes.[163]

In 1971, glutaraldehyde-treated bovine pericardial tissue (GTBPT) xenografts were used to replace defective human heart valves.[164] According to Silver and Levistsky,[17] GTBPT and GTPHV bioprostheses undergo an identical process of valve failure.

Heynan et al.[165] constructed valvular bioprostheses from autologous dog pericardial tissue treated with 87% glycerol for 5 min at 37°C. On implantation in dogs, the valves were covered with a thin layer of fibrin, including erythroblasts, erythrocytes, platelets, plasma cells, and/or immunoblast-like cells. According to them, the appearance of precocious but durable immunocompetent cells suggested that the glycerol treatment was responsible for the local immune reaction.[165] Glycerol is known to contain glutaraldehyde as a contaminant and probably was responsible for modifying the tissue.[15] Immunofluorescent reactions of plasma cells and valvular tissue with anti-dog immunoglobulin also indicated that the glycerol treatment of the autologous pericardium was responsible for inducing a temporary and local autoimmune disease.[165]

Antibovine heart valve sera has been shown to react with fibroblasts, endothelial cells, endocardium, plate of the valve, connective tissue of the arterial walls, reticulin around the muscle fibers of mitral valve, and myocardium of the human heart.[166,167]

Since GTBPT valves, like their porcine counterparts, have the same fate on implantation in humans,[17] we considered it necessary to determine whether GTBPT tissue xenografts were immunogenic or nonimmunogenic. Initial studies conducted with homogenates of GTBPT- and glutaraldehyde-treated bovine pericardial tissue stored in 4% formaldehyde (GTBPT) indicated that both of these tissues were capable of inducing an immune response.[140]

In a follow-up investigation, New Zealand white rabbits were grafted in the abdominal region with 1 × 2 cm strips (3 per rabbit) of UTBPT or GTBPT xenografts.[137-140] Immunogenic response to UTBPT and GTBPT xenografts was tested by indirect hemagglutination, leukocyte migration inhibition, and skin reaction tests.[140] Glutaraldehyde treatment inhibited the resorption of bovine pericardial tissue xenografts. Sera samples obtained from 5 of 8 UTBPT-implanted rabbits developed anti-UTBPT antibodies 30 days after primary graft implantation. Anti-UTBPT antibodies, obtained from 3 to 8 rabbits 90 days after initial surgery crossreacted with GTBPT antigens. Circulating antibodies against GTBPT antigens were detected in 6 of 8 GTBPT grafted rabbits' sera 30 and 60 days following surgery. Anti-GTBPT antibodies obtained 60 days after initial surgery from 7 of 8 rabbits crossreacted with UTBPT antigens.[137,140] In the primary xenograft implantation study, maximum anti-UTBPT titer ranges of 4 to 256 and anti-GTBPT titer ranges of 16 to 256 were observed 90 days after implantation of xenografts.[15,137,140]

Leukocyte migration inhibition (LIF) and skin reaction tests monitor cellular immune responses. Incubation of white blood cells with UTBPT (40.4 ± 4.5%) and GTBPT (26.6 ± 4.1%) antigens 90 days after the insertion of grafts significantly inhibited the migration of leukocytes obtained from UTBPT-grafted rabbits.[138,140] The migration of leukocytes obtained from GTBPT-rabbits was inhibited significantly on incubation with UTBPT antigens (42.5 ± 3.2%) after 60 days of implantation and with GTBPT homogenates 60 (34.6 ± 6.8%) and 90 days (53.8 ± 5.3%) after implantation of the tissues.[138,140] Positive skin responses were detected (90 days after initial surgery) in 7 of 8 UTBPT-treated rabbits and in 5 of 8 GTBPT-treated rabbits.[15,138,140]

The above investigation was extended by retrieving the primary grafts after 90 days of implantation and replacing them with secondary UTBPT or GTBPT xenografts.[139,140] The grafts were retrieved for morphologic examination 90 days after the secondary graft implantation and the animals were sacrificed by i.v. injection of concentrated sodium pento-

Table 1
## ANTIBODY TITERS OF SERA OBTAINED FROM SHAM-OPERATED CONTROLS AND RABBITS IMPLANTED WITH EITHER UTBPT OR GTBPT SECONDARY XENOGRAFTS

| Serum source (rabbits) | Rabbit No. | Days | | | | | |
|---|---|---|---|---|---|---|---|
| | | 30 | | 60 | | 90 | |
| | | UTBPT[a] | GTBPT[a] | UTBPT | GTBPT | UTBPT | GTBPT |
| Control | 8 | 0 | 0 | 0 | 0 | 0 | 0 |
| UTBPT/UTBPT | 4 | 16 | 0—4 | 16—64 | 0—4 | 16—64 | 0—4 |
| UTBPT/GTBPT | 4 | 16 | 0—4 | 4—64 | 0—4 | 16—64 | 0—4 |
| GTBPT/GTBPT | 4 | 4—16 | 64—256 | 0—16 | 64—256 | 16—64 | 64—256 |
| GTBPT/UTBPT | 4 | 4—16 | 256 | 4 | 256—1024 | 16 | 64—1024 |

*Note:* Antibody titers at 30, 60, and 90 days after implantation of secondary graft. Titers in hemagglutination units were read after 24 hr of incubation at 4°C.

[a]  Each sera sample was tested aganist UTBPT and GTBPT antigens by Indirect Hemagglutination Test.

From Salgaller, M. L., M.Sc. thesis, Universtiy of Dayton, Dayton, Ohio, 1981. With permission.

barbital. Implantation of UTBPT xenografts in previously GTBPT-implanted rabbits increased the titers of anti-GTBPT antibodies from 16 to 256 to 64 to 1024. Humoral responses were more specific to the particular or corresponding xenografts; e.g., when UTBPT was replaced with UTBPT xenograft and GTBPT was replaced with GTBPT xenograft (Table 1). Implantation of any type of secondary xenograft increased the degree of inhibition of leukocyte migration (Table 2).

After replacement surgery, rabbits implanted with corresponding primary and secondary tissue xenografts gave a positive skin reaction to UTBPT and GTBPT antigens. However, when UTBPT- and GTBPT-grafted rabbits were implanted with dissimilar GTBPT and UTBPT xenografts, the number of positive skin reactors decreased on intradermal challenge with either UTBPT or GTBPT antigens (Table 3).

All initial UTBPT grafts were resorbed by the host. All initial and secondary GTBPT grafts were retrieved for morphological examination. Gross morphological observation showed no distinct degeneration of GTBPT xenografts. Not all secondary UTBPT implants were resorbed. This suggested that the graft was protected by blocking antibodies.

Histopathological examination of GTBPT grafts recovered from the rabbits after 90 days of implantation showed DTH reaction. The graft site was infiltrated with small hyperchromatic lymphoid cells and some macrophages (Figure 1). The health of the animals was not impaired by the implantation of either untanned or tanned tissues in any of our investigations.

The data obtained in our investigations with tanned porcine heart valves and bovine pericardial tissue indicates that treatment of tissues with 0.5% glutaraldehyde does not render the tissue nonimmunogenic.

## V. CONCLUSIONS

Chemical and/or physical modifications of natural tissues in general modify and/or eliminate some antigenic determinants but do not reduce their immunologic potential. The loss of specificity towards the native antigenic determinants has often been interpreted as reduction in antigenicity.

At present, it is difficult to project the pathologic potential of the immune responses

**Table 2**
## PERCENT MIGRATION INHIBITION OF LEUKOCYTES OBTAINED FROM RABBITS AFTER IMPLANTATION OF DIALYSIS TUBING AND SECONDARY GRAFTS

| Lymphocyte source (rabbits) | No. of rabbits | Days of secondary graft implantation | % Migration inhibition ± SEM in the presence of | |
|---|---|---|---|---|
| | | | UTBPT antigen | GTBPT antigen |
| Control | 8 | 30 | 0.0 ± 0.0 | 0.0 ± 0.0 |
| | 8 | 60 | 0.0 ± 0.0 | 0.0 ± 0.0 |
| | 7 | 90 | 0.0 ± 0.0 | 0.0 ± 0.0 |
| UTBPT/UTBPT | (4) | 30 | 43.6 ± 3.0[a] | 38.5 ± 3.4[a] |
| | 4 | 60 | 48.5 ± 3.4[a] | 43.4 ± 2.7[a] |
| | 4 | 90 | 58.3 ± 1.3[a] | 50.0 ± 4.6[a] |
| UTBPT/GTBPT | 4 | 30 | 38.1 ± 6.4[a] | 25.3 ± 5.4[a] |
| | 4 | 60 | 51.3 ± 6.7[a] | 60.2 ± 3.2[a] |
| | 4 | 90 | 56.7 ± 3.6[a] | 48.9 ± 6.0[a] |
| GTBPT/GTBPT | 4 | 30 | 48.0 ± 3.3[a] | 48.8 ± 7.4[a] |
| | 4 | 60 | 45.9 ± 3.2[a] | 53.9 ± 4.4[a] |
| | 4 | 90 | 52.3 ± 1.2[a] | 53.4 ± 3.0[a] |
| GTBPT/UTBPT | 4 | 30 | 49.4 ± 4.3[a] | 58.6 ± 5.2[a] |
| | 4 | 60 | 61.4 ± 4.0[a] | 62.3 ± 3.1[a] |
| | 4 | 90 | 64.3 ± 1.4[a] | 67.9 ± 1.8[a] |

*Note:* Lymphocytes obtained from rabbits grafted with dialysis tubing, untreated (UTBPT), or glutaraldehyde-treated bovine pericardial tissue (GTBPT) were tested for LIF activity with UTBPT and GTBPT antigens.

[a]  Significant at $p < 0.05$

From Salgaller, M. L.,M.Sc.  thesis, University of Dayton, Dayton, Ohio, 1981. With permission.

**Table 3**
## SKIN REACTIONS OF RABBITS IMPLANTED WITH DIALYSIS TUBING, SECONDARY UNTREATED (UTBPT), OR GTBPT XENOGRAFTS 48 HR AFTER I.D. CHALLENGE WITH UTBPT OR GTBPT HOMOGENATES

| Rabbits implanted with | Days of implantation | UTBPT antigen | | GTBPT antigen | |
|---|---|---|---|---|---|
| | | No. of positive reactors | Induration diameter (mm) | No. of positive reactors | Induration diameter (mm) |
| Dialysis | 60 | 0/4 | — | 0/4 | — |
| tubing | 90 | 0/4 | — | 0/4 | — |
| UTBPT/UTBPT | 60 | 4/4 | 5.9 ± 0.8[a] | — | — |
| xenograft | 90 | 3/4 | 5.6 ± 0.2[a] | 3/4 | 5.8 ± 0.5[a] |
| UTBPT/GTBPT | 60 | — | — | 0/4 | — |
| xenograft | 90 | 0/4 | — | 0/4 | — |
| GTBPT/GTBPT | 60 | — | — | 0/4 | — |
| xenograft | 90 | 1/4 | 5.0 ± 0.0[a] | 3/4 | 5.2 ± 0.1[a] |
| GTBPT/UTBPT | 60 | 0/4 | — | — | — |
| xenograft | 90 | 0/4 | — | 1/4 | 5.5 ± 0.0[a] |

[a]  Mean ± SEM (Indurations of 5 mm or more are scored as positive skin reactions).

From Salgaller, M. L., M.Sc. thesis, University of Dayton, Dayton, Ohio, 1981. With permission.

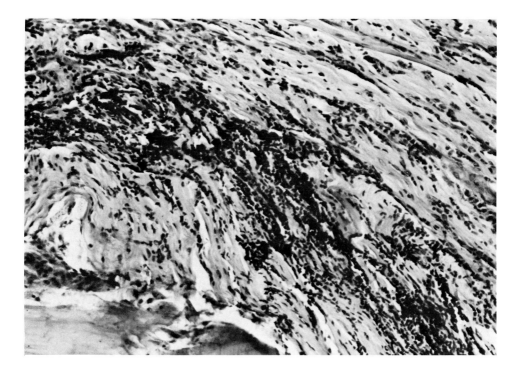

FIGURE 1. Photomicrograph of a section of GTBPT implanted in rabbit for 90 days showing macrophages, diffused infiltration of small hyperchromatic lymphoid cells, and delayed hypersensitivity reaction.

generated by the components of various bioprostheses. However, it seems that in some instances generation of an immune response is beneficial to the bioprosthesis.

## ACKNOWLEDGMENTS

The author thanks Dr. C. Parkash, Department of Microbiology and Immunology, The Ohio State University, Columbus, Ohio for reviewing the manuscript, Dr. R. A. Solomon of Wright Patterson Air Force Base, Dayton, Ohio for interpreting micrographs of tissue sections, Robin A. Bajpai for editing the manuscript, and Ann Feldmann for typing this manuscript.

## REFERENCES

1. **Moon, D. C.,** Long term effects of prosthetic materials, *Am. J. Cardiol.,* 50, 621, 1982.
2. **Clark, W. R.,** *The experimental Foundations of Modern Immunology,* 2nd ed., John Wiley & Sons, N.Y., 1983, chaps. 1, 5 to 7 and 10.
3. **Kimball, J. W.,** *Introduction to Immunology,* MacMillan, N.Y., 1983, chaps. 1, 3, and 13.
4. **McConnell, I., Munroe, A., and Waldmann, H.,** A course on the molecular and cellular basis of immunity, in *The Immune System,* 2nd ed., Blackwell Scientific, Oxford, 1981, chaps. 6 and 15 to 16.
5. **Eisen, H. N.,** *Immunology,* 2nd ed., Harper & Row, Philadelphia, 1980, chaps. 15 and 18.
6. **Hokama, Y. and Nakamura, R. M.,** *Immunology and Immunopathology. Basic Concepts,* Little, Brown Boston, 1982, chaps. 3 to 4.

7. **'s-Gravenmade, E. J. and Wemes, J. C.,** Interaction of glutaraldehyde with biological materials in endodontics, *J. Dent. Res.,* 52, 501, 1973.

8. **Carpentier, A., Lemaigre, G., Robert L., Carpentier, S., and Dubost, C.,** Biological factors affecting long term results of valvular grafts, *J. Thorac. Cardiovasc. Surg.,* 58, 467, 1969.

9. **Richard, F. and Knowles, F.,** Glutaraldehyde as a protein cross-linking reagent, *J. Mol. Biol.,* 37, 231, 1968.

10. **Habeeb, A. and Hiramoto, R.,** Reaction of proteins with glutaraldehyde, *Arch. Biochem. Biophys.,* 125, 16, 1968.

11. **Cunningham, K. W., Lazzari, E. P., and Ranly, D. M.,** The effect of form-cresol and glutaraldehyde on certain enzymes in bovine dental pulp, *Oral. Surg.,* 54, 100, 1982.

12. **Bajpai, P. K., Russo, D. A., Slanczka, D. J., and Weiskettel, L. E.,** Immunogenicity of glutaraldehyde-treated tissues used in heart valve replacement, in *Advances in Biomaterials,* Vol. 3, Winters, G. D., Gibbons, D. F., and Plenk, H., Eds., John Wiley & Sons, Chicester, U.K. 1982, 243.

13. **Bajpai, P. K.,** Antigenicity of glutaraldehyde stabilized biological materials, in *Biomaterials in Reconstructive Surgery,* Rubin, L. R., Ed., C. V. Mosby, St. Louis, 1983, chap. 17.

14. **Bajpai, P. K.,** Immunogenicity of tanned tissues used in valve xenografts, in *Biomaterials in Reconstructive Surgery,* Rubin, L. R., Ed., C. V. Mosby, St. Louis, 1983, chap. 47.

15. **Bajpai, P. K. and Salgaller, M. L.,** Immune responses to glutaraldehyde-treated xenografts, in *Biocompatible Polymers, Metals and Composites,* Szycher, M., Ed., Technomic Publ., Lancaster, Pa., 1983, chap. 17.

16. **Carpentier, A.,** From valvular xenograft to valvular bioprosthesis (1965—1977), *J. Assoc. Adv. Med. Instrum.,* 11, 98, 1977.

17. **Silverman, N. A. and Levitsky, S.,** Current choices for prosthetic valve replacement, *Mod. Concepts Cardiovasc. Dis.,* 52, 35, 1983.

18. **Peracchia, C. and Mittler, B.,** New glutaraldehyde fixation procedures, *J. Utlrastruct. Res.,* 39, 57, 1972.

19. **Eckert, B. S. and Snyder, J. A.,** Combined immunofluorescence and high voltage electron microscopy of cultured mammalian cells, using an antibody that binds to glutaraldehyde-treated tubulin, *Proc. Natl. Acad. Sci. U.S.A.,* 75, 334, 1978.

20. **Avrameas, S., Ternynck, T., and Guedson, J. L.,** Coupling of enzymes to antibodies and antigens, *Scand. J. Immunol.,* 8(7), 7, 1978.

21. **Leshem, B. and Naor, D.,** Studies on the immune responses to fixed antigens. IV. Recall of immunologic memory and antibody production, *Immunology,* 36, 775, 1979.

22. **Stanley, W. L., Watters, G. G., Kelly, S. H., and Olson, A. C.,** Glucomylase immobilized on chitin with glutaraldehyde, *Biotechnol. Bioeng.,* 20, 135, 1978.

23. **Woodroof, E. A.,** Use of glutaraldehyde and formaldehyde to process tissue heart valves, *J. Bioeng.,* 2, 1, 1978.

24. **Knapp, T. R.,** Development of an injectable collagen for soft-tissue restoration, in *Biomaterials in Reconstructive Surgery,* Rubin, L. R., Ed., C. V. Mosby, St. Louis, 1983, chap. 58.

25. **Simpson, R. L.,** Collagen as a biomaterial, in *Biomaterials in Reconstructive Surgery,* Rubin, L. R., Ed., C. V. Mosby, St. Louis, 1983, chap. 11.

26. **Bigley, N. J., Kreps, D. B., Smith, R. A., and Esa, A.,** Antigenic modification, rosette forming cells, and *Salmonella typhimurium* resistance in outbred and inbred mice, *Infect. Immun.,* 32, 353, 1981.

27. **Patterson, R., Suzsko, I. M., Leiss, C. R., Pruzansky, J. J., and Bacal, E.,** Comparison of immune reactivity to polyvalent monomeric and polymeric ragweed antigens, *J. Allergy Clin. Immunol.,* 61, 28, 1978.

28. **Ternynck, T. and Avrameas, S.,** Polymerization and immobilization of proteins using ethylchloroformate and glutaraldehyde, *Scand. J. Immunol. Suppl.,* 3, 39, 1976.

29. **Frost, P. and Sanderson, C. J.,** Tumor immunoprophylaxis in mice using glutaraldehyde-treated syngeneic tumor cells, *Cancer Res.,* 35, 2646, 1975.

30. **Frost, P., Edwards, A., and Sanderson, C. J.,** The use of glutaraldehyde fixation for the study of the immune response to syngeneic tumor antigen [discussion paper], *Ann. N.Y. Acad. Sci.,* 276, 98, 1976.

31. **Broom, N. D.,** Fatigue-induced damage in glutaraldehyde preserved heart valve tissue, *J. Thorac. Cardiovasc. Surg.,* 76, 202, 1978.

32. **Schechter, I., Belledegrin, A., Ben-Basat, M., and Kaplan, I.,** Prolonged retention of glutaraldehyde-treated skin homografts in humans, *Br. J. Plast. Surg.,* 28, 198, 1975.

33. **Silver, M. D.,** Late complications of prosthetic heart valves, *Arch. Pathol. Lab. Med.,* 102, 281, 1978.

34. **Johansson, S. G. O., Deuschl, H., and Letterstrom, O.,** Use of glutaraldehyde modified timothy grass pollen extract in nasal hyposensitization of hay fever, *Int. Arch. Allergy Appl. Immunol.,* 60, 447, 1979.

35. **Habal, M. B., Maniscola, J. E., and Leake, D. C.,** Implantology for corrections of cranio-facial defects with and elastomer composite and bone, *Biomater. Med. Devices Artif.Organs,* 7, 271, 1979.

36. **Lane, S. L.,** Plastic procedures as applied to oral surgery, *J. Oral. Surg.,* 16, 489, 1958.
37. **Marciani, R. D., Gonty, A. A., Giansanti, J. S., and Avilla, J.,** Autogenous cancellous-marrow bone grafts in irradiated dog mandibles, *Oral. Surg.,* 43, 365, 1977.
38. **Naber, C. L., Reid, O. M., and Hamner, J. E., III,** Gross and histologic evaluation of an autogeneous bone graft 57 months postoperatively, *J. Periodontol.,* 43, 702, 1972.
39. **Bajpai, P. K.,** Biodegradable scaffolds in orthopedic, oral and maxillofacial surgery, in *Biomaterials in Reconstructive Surgery,* Rubin, L. R., Ed., C. V. Mosby, St. Louis, 1983, chap.22.
40. **Mankin, H. J., Fogelson, F. S., Thresher, A. Z., and Jaffer, F.,** Massive reaction and allograft transplantation in the treatment of malignant bone tumors, *N. Engl. J. Med.,* 294, 1247, 1976.
41. **Nade, S.,** Osteogenesis after bone and bone marrow transplantation, *Acta Orthop. Scand.,* 48, 572, 1977.
42. **Habal, M. D.,** Current status of biomaterials. Clinical applications in plastic and reconstructive surgery, *Biomater. Med. Devices Artif. Organs,* 7, 229, 1979.
43. **Zins, J. E. and Whitaker, L. A.,** Membrane versus endochondral bone autografts implications for craniofacial reconstruction, *Surg. Forum,* 30, 521, 1979.
44. **Cloward, R. B.,** Gas sterilized cadaver bone grafts for spinal fusion operations. A simplified bone bank, *Spine,* 5, 4, 1980.
45. **Nade, S.,** Clinical implications of cell function in osteogenesis. A reappraisal of bone graft surgery, *Am. R. Coll. Surg.,* 61, 189, 1979.
46. **Golan, J., Shapira, Y., Ben Hur, N., and Dollberg, L.,** Bone formation from periosteal grafts and investigation on the possible effects of calcitonin, *J. Surg. Res.,* 21, 339, 1976.
47. **Herndon, C. H. and Chase, S. N.,** Experimental studies in transplantation of whole joints, *J. Bone Jt. Surg.,* 34A, 339, 1976.
48. **Goncalvez, R. J. and Merzel, J.,** Homogenous and heterogenous bone implants preserved by formaldehyde a histologic study, *J. Am. Dent. Assoc.,* 93, 1165, 1976.
49. **Freeman, E. and Turnbull, R. S.,** Histological evaluation of freeze-dried fine particle bone allografts, *J.Periodontol.,* 48, 288, 1977.
50. **Langer, F., Czitron, A., Pritzker, K. P., and Gross, A. E.,** The immunogenicity of fresh and frozen allogeneic bone, *J. Bone Jt. Surg.,* 57A, 216, 1975.
51. **Muscolo, D. L., Kawai, S., and Ray, R. D.,** Cellular and humoral immune response analysis of bone allografted rats, *J. Bone Jt. Surg.,* 58A, 326, 1976.
52. **Nisbet, N. W.,** Antigenicity of bone, *J. Bone Jt. Surg.,* 59B, 263, 1977.
53. **Burwell, R. G.,** The fate of freeze-dried bone allografts, *Transplant. Proc.,* 8, 95, 1976.
54. **Oikarinen, J. and Kornhonen, L. K.,** The bone inductive capacity of various bone transplanting materials used for treatment of experimental bone defects, *Clin. Orthop.,* 140, 208, 1979.
55. **Bauermeister, A.,** Formation, establishment and result of Kieler's protein-free animal bone chip grafts, *Bull. Soc. Int. Chir.,* 19, 612, 1980.
56. **Haasch, K.,** Clinical experiences with the carinated graft, *Chirurg,* 34, 21, 1963.
57. **Kosch, W. and Dahmer, G.,** Experimental and clinical experiences with the heterologous bone bank graft, *J. Orthop.,* 96, 348, 1962.
58. **Lubinas, H.,** Area of use of the Kiel bone graft, *Med. Pharm. Mitteil.,* 100, 2504, 1963.
59. **Maatz, R. and Bauermeister, A.,** Essential features and applications of the Kiel bone graft, *Med. Pharm. Mitteil.,* 97, 2117, 1961.
60. **Taheri, Z. E. and Gueramy, M.,** Experience with calf bone in cervical interbody spinal fusions, *J. Neurosurg.,* 36, 67, 1972.
61. **Boyne, P. J. and Cooksey, D. E.,** Use of cartilage and bone implants in restoration of edentulous ridges, *J. Am. Dent. Assoc.,* 71, 1426, 1965.
62. **Heyner, S.,** The antigenicity of cartilage grafts, *Surg. Gynecol. Obstet.,* 136, 298, 1973.
63. **Lye, T. L.,** A histologic evaluation of cartilage homograft implant used in prosthetic surgery, *Oral Surg.,* 31, 745, 1971.
64. **Meijer, R. and Walia, I. S.,** Preserved cartilage to fill facial bone defects, in *Biomaterials in Reconstructive Surgery,* Rubine, L. R., Ed., C. V. Mosby St Louis, 1983, chap. 32.
65. **Smith, A.,** Survival of frozen chondrocytes isolated from cartilage of adult animals, *Nature(London),* 205, 782, 1965.
66. **Gillies, H. D. and Kistensen, H. K.,** Ox cartilage in plastic surgery, *Br. J. Plast. Surg.,* 4, 63, 1951.
67. **Gibson, T., Davis, W. B., and Gillies, H. D.,** The encapsulation of preserved cartilage grafts with prolonged survival, *Br. J. Plast. Surg.,* 12, 22, 1959.
68. **deCarvalho, A. C. P. and Okamoto, T.,** Autogenous and homogenous cartilage grafts. I. Implantation into subcutaneous connective tissue of rats, *Rev. Fac. Odont. Aracatuba,* 6, 13, 1977.
69. **deCarvalho, A. C. P. and Okamoto, T.,** A histologic study of cartilage autografts and allografts placed in dental sockets of rats, *J. Oral Surg.,* 37, 11, 1979.
70. **Peer, L. A.,** *Transplantation of Tissues,* Vol. 1, Williams & Wilkins, Baltimore, 1955, 110.

71. **Senn, N.,** On the healing of aseptic bone cavities by implantation of antiseptic decalcified bone, *Am. J. Med. Sci.,* 98, 219, 1889.

72. **Glant, T., Hadhazy, C. S., Bordan, L., and Hasmati, S.,** Antigenicity of bone tissue. I. Immunological and immunohistochemical study of non-collagenous proteins of the bovine cortical bone, *Acta Morphol. Acad. Sci. Hung.,* 23, 111, 1975.

73. **Narang, R., Lloyd, W., and Wells, H.,** Grafts of decalcified allogeneic bone matrix promote the healing of fibular fracture gaps in rats, *Clin. Orthop.,* 80, 174, 1971.

74. **Narang, R., Wells, H., and Laskin, D. M.,** Ridge augmentation with decalcified allogeneic bone matrix graft in dogs, *J. Oral. Surg.,* 30, 722, 1972.

75. **Reddi, A. H. and Huggins, C. B.,** Influence of geometry of transplanted tooth and bone on transformation of fibroblasts, *Proc. Soc. Exp. Biol. Med.,* 143, 634, 1973.

76. **Register, A. A., Scopp, I. W., Kassouny, D. Y., Pfau, F. R., and Peskin, D.,** Human bone induction by allogeneic dentin matrix, *J.Peridodontol.* 43, 459, 1979.

77. **Rüedi, T. P. and Luscher, J. N.,** Results after internal fixation of comminuted fractures of the femoral shaft with D.C. plates, *Clin. Orthop.,* 138, 74, 1979.

78. **Sharrard, W. J. and Collins, D. H.,** The fate of human decalcified bone grafts, *Proc. Soc. Med.,* 54, 1101, 1961.

79. **Tornberg, D. N. and Bassett, C. A. L.,** Activation of the resting periosteum, *Clin. Orthop.,* 129, 305, 1977.

80. **Urist, M. R., Earnest, F., Iv., Kimball, D. M., DiJullio, T. P., and Iwata, H.,** Bone morphogenesis in implants of residues and radioisotope-labelled bone matrix, *Calcif. Tissue Res.,* 15, 269, 1974.

81. **Urist, M. R., Mikulski, A. J., and Boyd, S. D.,** A chemosterilized antigen-extracted autodigested allo-implant for bone banks, *Arch.Surg.,* 110, 416, 1975.

82. **Joos, U., Ohcs, G., and Ries, P. E.,** Influence of collagen fleece on bone regeneration, *Biomaterials,* 1, 23, 1981.

83. **Collins, D., Lindberg, K., McLees, B., and Pinnell, S.,** The collagen of heart valve, *Biochim. Biophys. Acta,* 495, 129, 1977.

84. **Jimenez, S. A. and Bashey, R. I.,** Solubilization of bovine heart valve collagen, *Biochem. J.,* 173, 337, 1978.

85. **Drake, M. P., Davidson, P. F., Bumps, S., and Schmitt, F. O.,** Action of proteolytic enzymes on tropocollagen and insoluble collagen, *Biochemistry,* 5, 301, 1966.

86. **Rothland, S. and Watson, R.,** Immunologic reaction among various animal collagens, *J. Exp. Med.,* 122, 441, 1965.

87. **Trentham, D. E., Townes, A. S., and King, A. H.,** Autoimmunity to type II collagen: an experimental model of arthritis, *J. Exp. Med.,* 146, 857, 1977.

88. **Trentham, D. E., Townes, A. S., King, A. H., and David, J. R.,** Humoral and cellular sensitivity to collagen in type II collagen-induced arthritis in rats, *J. Clin. Invest.,* 61, 89, 1978.

89. **Adelmann, B.,** The structural basis of cell-mediated immunological reaction of collagen, *Immunology,* 23, 739, 1972.

90. **Shakespeare, P. G. and Griffiths, R. W.,** Dermal collagen implants in man, *Lancet,* 1, 795, 1980.

91. **Pullinger, B. D. and Pirie, A.,** Chronic inflammation due to implanted collagen, *J. Pathol. Bacterial.,* 54, 341, 1942.

92. **Menzel, J.,** Radioimmunoassay for anti-collagen antibodies using $^{14}$C-labelled collagen, *J. Immunol. Methods,* 15, 77, 1977.

93. **Stenzel, K. H., Dunn, M. W., and Rubin, A. L.,** Collagen gels and designs for a vitreous replacement, *Science,* 164, 1282, 1978.

94. **Cochran, M. D. and Hait, M. R.,** An experimental study on healing of bone following application of microcrystalline collagen hemostatic agent, *J. Trauma,* 15, 404, 1975.

95. **Chvapil, M.,** Collagen sponge: theory and practice of medical applications, *J. Biomed. Mater. Res.,* 11, 721, 1977.

96. **Speer, D., Chvapil, M., Volz, R., and Holme, M.,** Enhancement of healing in osteochondral defects by collagen sponge implants, *Clin. Orthop.,* 144, 326, 1979.

97. **Moskow, B. S., Karsh, F., and Skin, S. D.,** Histological assessment of autogenous bone autograft, *J. Periodontal.,* 50, 291, 1979.

98. **Delbaso, A. M. and Adrian, J. C.,** Collagen gel in osseous defects, *Oral Surg.,* 42, 562, 1976.

99. **Schmitt, F. O., Levine, L., Drake, M. P., Rubin, A. L., Pfahl, D., and Davison, P. F.,** The antigenicity of tropocollagen, *Prod. Natl. Acad. Sci. U.S.A.,* 51, 493, 1964.

100. **Davison, P. F., Levine, L., Drake, M. P., Rubin, A. L., and Bump, S.,** The serologic specificity of tropocollagen telopeptides, *J. Exp. Med.,* 126, 331, 1967.

101. **Makkes, P. C., van Velzen, S. K., and van den Hooff, A.,** The response of the living organism to dead and fixed dead enclosed homologous tissue, *Oral Surg.,* 136, 296, 1978.

102. **Makkes, P. C., van Velzen, S. K., and van den Hooff, A.,** Short-term tissue reactions to enclosed glutaraldehyde-fixed tissue, *Oral Surg.,* 46, 854, 1978.

103. **Habeeb, A.,** A study of the antigenicity of formaldehyde and glutaraldehyde-treated bovine serum albumin and albumin-bovine serum albumin conjugate, *J. Immunol.,* 102, 457, 1969.

104. **Capperauld, I.,** Bovine fibrin implants: tissue reaction, in *Biomaterials in Reconstructive Surgery,* Rubin, L. R., Ed., C. V. Mosby St. Louis, 1983, chap.

105. **Horvath, N., Somogyvari, K., Sormos, P., and Surjan, J.,** Experimental studies on the antigenic action of the Bioplast preparations, *Acta Vet. Acad. Sci. Hung.,* 19, 191, 1969.

106. **Salisbury, R. E.,** Synthetic skin and skin substitutes, in *Biomaterials in Reconstructive Surgery,* Rubin, L. R., Ed., C. V. Mosby St. Louis, 1983, chap. 54.

107. **Harris, N., Compton, J., Abston, S., and Larson, D.,** Comparison of fresh, frozen and lyophilized porcine skin as xenografts on burned patients, *Burns,* 2, 71, 1976.

108. **Hiratzka, L. F. and Wright, C. B.,** Experimental and clinical results of grafts in venous system, a current review, *J. Surg. Res.,* 25, 542, 1978.

109. **Rosenberg, N.,** The bovine arterial graft and its several applications, *Surg. Gynecol. Obstet.,* 142, 104, 1976.

110. **Perloff, L. T., Reckard, C. R., Rowlands, D. T., and Barker, C. F.,** The venous homograft, an immunological question, *Surgery,* 72, 961, 1972.

111. **Schwartz, S. I., Kutner, F. R., Neistadt, A., Basner, H., Resnicoff, S., and Vangdon, J.,** Antigenicity of homografted veins, *Surgery,* 61, 471, 1967.

112. **Williams, G, M., Hoar, A., Krajewski, C., Parks, L. C., and Roth, J.,** Rejection and repair of endothelium in major vessel transplants, *Surgery,* 78, 694, 1975.

113. **Shiel, A. G. R., Stephan, M. S., Boulas, J., Johnson, D. S., and Lowenthal, J.,** Small arterial reconstruction using modified cadaveric saphenous veins, *Am. J. Surg.,* 134, 591, 1977.

114. **Esato, K., Shintani, K., and Yasutake, S.,** Modification and morphology of human umbilical cord vein as canine arterial bypass grafts, *Ann. Surg.,* 191, 443, 1980.

115. **Wahl, S. M. and Wahl, L. M.,** Lymphokine modulation of connective tissue metabolism, *Ann. N.Y. Acad. Sci.,* 332, 441, 1979.

116. **Defalco, R. J.,** Immunologic studies of untreated and chemically modified bovine carotid arteries, *J. Surg. Res.,* 10, 95, 1970.

117. **Mattila, S. O. and Fogarty, T. J.,** Antigenicity of vascular heterografts, *Surg. Res.,* 15, 18, 1973.

118. **Frolova, M., Barbarash, L., Gudkova, R., and Karpinskaya, J.,** Effect of method of conservation on immunogenicity and antigenic composition of xenograft heart valve tissues, *Byull. Eksp. Biol. Med.,* 73, 306, 1973.

119. **Sheikh, K. M., Tascon, M. A., and Nimni, M. E.,** Autoimmunity in patients with Hancock valve implant and bypass heart surgery, *Fed. Proc. Am. Soc. Exp. Biol.,* 39, 472, 1980.

120. **Dale, N. A. and Lewis, M. R.,** Further experience with bovine arterial grafts, *Surgery,* 80, 711, 1976.

121. **Okamura, K., Chiba, C., Iriymam, T., Itoh, T., Maeta, H., Ijima, H., Mitsui, T., and Hori, M.,** Antigen depressant effect of glutaraldehyde for aortic heterografts with a valve, with special reference to a concentration right fit for the preservation of grafts, *Surgery,* 87, 170, 1980.

122. **Russo, D. R.,** Studies on the Antigenicity of Porcine Intima, M. Sc. thesis, University of Dayton, Dayton, Ohio, 1979.

123. **Baig, M., Diacoff, G., and Ayoub, E.,** Comparative studies of the acid mucopolysaccharide composition of rheumatic and normal heart valves in man, *Cir. Res.,* 42, 271, 1978.

124. **Barratt-Boyes, B. G.,** Cardiothoracic surgery in the antipodes, *J. Thorac. Cardiovasc. Surg.,* 78, 804, 1979.

125. **Kirklin, J. W.,** The replacement of cardiac valves, *N. Engl. J. Med.,* 30, 291, 1981.

126. **Thiede, A., Timm, C., Bernhard, A., and Miller-Ruchholtz, W.,** Studies on the antigenticity of vital allogeneic valve leaflet transplants in immunologically controlled strain combinations, *Transplantation,* 26, 391, 1978.

127. **Lefrak, A. E. and Starr, A.,** *Cardiac Valve Prosthesis,* Appleton-Century-Crofts, N.Y., 1979, 301.

128. **Stephens, B. J. and O'Brien, M. F.,** Pathology of xenografts in aortic valve replacement, *Pathology,* 4, 167, 1972.

129. **Davey, T., Kaufman, B., Smelloff, E., and Cartwright, R.,** Design of heart valve, *Mech. Eng.,* 7, 22, 1966.

130. **Spray, L. T. and Roberts, W. C.,** Structural changes in porcine xenografts used as substitute cardiac valves, *Am. J. Cardiol.,* 40, 19, 1977.

131. **Spray, T. L. and Roberts, W. C.,** Structural changes in Hancock porcine xenograft cardiac valve bioprosthesis, *Adv. Cardiol.,* 22, 241, 1978.

132. **Korngold, R. and Sprent, J.,** Selection of cytotoxic T-cell precursors specific for minor histocompatibility determinants, *J. Exp. Med.,* 151, 314, 1980.

133. **Slanczka, D. J. and Bajpai, P. K.,** Immunogenicity of glutaraldehyde-treated porcine heart valves, *Int. Res. Commun. Syst. Med. Sci.,* 6, 421, 1978.

134. **Slanczka, D. J., Russo, D. A., Bajpai, P. K., and Woodroof, E. A.,** Immunogenicity of tanned tissue used in heart valve replacement, in *Proceedings of the Seventh New England (Northeast) Bioengineering Conference,* Welkowitz, W., Ed., Pergamon Press, Oxford, U.K., 1979. 197.

135. **Bajpai, P. K., Stull, P. A., and Anderson, J. M.,** Immunogenicity of glutaraldehyde cross-linked porcine heart valve xenografts, *Int. Res. Commun. Syst. Med. Sci.,* 8, 519, 1980.

136. **Bajpai, P. K. and Stull, P. A.,** Glutaraldehyde cross-linked porcine heart valve xenografts and cell-medicated immune response, *Int. Res. Commun. Syst. Med. Sci.,* 8, 642, 1980.

137. **Salgaller, M. L. and Bajpai, P. K.,** The effect of glutaraldehyde-treated bovine pericardial tissue xenografts on humoral immune responses in rabbits, *Int. Res. Commun. Syst. Med. Sci.,* 9, 672, 1981.

138. **Salgaller, M. L. and Bajpai, P. K.,** Glutaraldehyde-treated bovine pericardial tissue xenografts and cell-medicated immune responses, *Int. Res. Syst. Med. Sci.,* 9, 868, 1981.

139. **Salgaller, M. L. and Bajpai, P. K.** Immunogenicity of glutaraldehyde-treated bovine pericardial tissue xenografts, *Transac. 8th Annu. Meet. Soc. Biomater.,* 5, 125, 1983.

140. **Salgaller, M. L.,** Immunogenicity of Glutaraldehyde-Treated Bovine Pericardial Tissue Xenografts, M.Sc. thesis, University of Dayton, Dayton, Ohio, 1981.

141. **Ferrans, V. J., Spray, L. T., Billingham, M. E., and Roberts, W. C.,** Structural changes in glutaraldehyde-treated porcine heterografts used as substitute cardiac valves, *Am. J. Cardiol.,* 48, 1159, 1978.

142. **Ferrans, V. J., Spray, T. L., Billingham, M. E., and Roberts, W. C.,** Ultrastructure of Hancock porcine valvular heterografts, *Circulation,* 58, 10, 1978.

143. **Rocchini, A. P., Weesner, K. M., Heidelberger, K., Keren, O., Behrendt, D., and Rosenthal, K.,** Porcine xenograft failure in children: an immunologic response, *Circulation,* 64, 162, 1981.

144. **Geha, A. S., Laks, H., Stansel, H. C., Cornhill, J. F., Killman, J. W., Buckley, M. J., and Roberts, W. C.,** Late failure of porcine valve heterografts in children, *J. Thorac. Cardiovasc. Surg.,* 78, 351, 1979.

145. **Thandroyen, F. T., Whitton, I. N., Pirie, D., Rogers, M. A., and Mitha, A. S.,** Severe calcification of glutaraldehyde-preserved porcine xenografts in children, *Am. J. Cardiol.,* 45, 690, 1980.

146. **Fiddler, G. I., Gerlis, L. M., Walker, D. R., Scott, O., and William G. J.,** Calcification of glutaraldehyde-preserved porcine and bovine xenograft valves in young children, *Ann. Thorac. Surg.,* 35, 257, 1983.

147. **Curcio, C. A., Commerford, P. J., Rose, A. G., Stevens, J. E., and Barnard, M. S.,** Calcification of glutaraldehyde-preserved porcine xenografts in young patients, *J. Thorac. Cardiovasc. Surg.,* 81, 621, 1981.

148. **Ashraf, M. and Bloor, C. M.,** Structural alterations of the porcine heterograft after various durations of implantations, *Am. J. Cardiol.,* 41, 1185, 1978.

149. **Brown, J. W., Dunn, M. M., Spooner, E., and Kirsh, M.,** Late spontaneous disruption of a porcine xenograft mitral valve, *J. Thorac. Cardiovasc. Surg.,* 75, 606, 1978.

150. **Ferrans, V. J., Boyce, S. W., Billingham, M. E., Jones, M., Ishihara, T., and Roberts, W. C.,** Calcific deposits in porcine bioprosthesis. Structure and pathogenesis, *Am. J. Cardiol.,* 46, 721, 1980.

151. **Fishbein, M. D., Gissen, S., Collins, J., Barsamian, E., and Cohn, L.,** Pathologic findings after cardiac valve replacement with glutaraldehyde-fixed porcine valves, *Am. J. Cardiol.,* 40, 331, 1977.

152. **Levy, R. J.,** The pathogenesis of porcine xenograft valve calcification, in *Biocompatible Polymers, Metals and Composites,* Szycher, M., Ed., Technomic Publ., Lancaster, Pa., 1983, chap. 18.

153. **Grehl, T. M., Oniel, M. B., Jr., Naifeh, J., Dajee, A., Hurley, E. J., and Riemenschneider, T.,** Spontaneous heterograft aortic valve failure, *Ann. Thorac. Surg.,* 31, 274, 1981.

154. **Ferrans, V. J., McManus, B., and Roberts, W. C.,** Cholesteryl ester crystals in a porcine aortic valve valvular bioprosthesis implanted for eight years, *Chest,* 83, 698, 1983.

155. **Valente, M., Bortolotti, U., Arbustini, E., Talenti, E., Thiene, G., and Galluci, V.,** Glutaraldehyde-preserved porcine bioprosthesis. Factors affecting performance as determined by pathologic studies, *Chest,* 83, 607, 1983.

156. **Frolova, M. A., Danilova, T. A., Gudkova, R. G., Bukova, V. A., and Fursov, B. A.,** Immunological indices in acquired heart defects after heart valve replacement with bioprostheses, *Kardiologia,* 21, 86, 1981.

157. **MaGilligan, D. J., Jr., Lewis, J. W., Jr., Heinzerling, R. H., and Smith, D.,** Fate of a second porcine bioprosthetic valve, *J. Thorac. Cardiovasc. Surg.,* 85, 362, 1983.

158. **Camilleri, J. P., Pornin, B., and Carpentier, A.,** Structural changes of glutaraldehyde-treated porcine bioprosthetic valves, *Arch. Pathol. Lab. Med.,* 106, 490, 1982.

159. **Villa, M. L., DeBiasi, S., and Pilotto, F.,** Residual heteroantigenicity of glutaraldehyde-treated porcine cardiac valves, *Tissue Antigens,* 16, 62, 1980.

160. **Ramos, A., Zavala, F., and Hoecker, G.,** Immune response to glutaraldehyde-treated cells. I. Dissociation of immunological memory and antibody production, *Immunology,* 36, 775, 1979.

161. **Kahan, M., Berman-Goldman, R., Saltoun, R., and Naor, D.,** Studies on the immune response to fixed antigens. Preferential induction of helper function with heavily trinitrophenylated sheep erythrocytes and glutaraldehyde-treated sheep erythrocytes, *J. Immunol.,* 117, 16, 1976.

162. **Tomecki, J.,** The influence of immunization of syrian hamsters with tumor cells treated with glutaraldehyde on transplantation immunity and the cytotoxic effect of lymphocytes on polyoma tumor cells, *Arch. Immunol. Ther. Exp.,* 27, 209, 1979.

163. **van Noort, R., Yates, S. P., Martin, T. R. P., Barker, A. T., and Black, M. M.,** A study of the effects of glutaraldehyde and formaldehyde on the mechanical behaviour of bovine pericardium, *Biomaterials,* 3, 21, 1982.

164. **Ionescu, M. I., Tandon, A. P., Mary, D. A. S., and Abid, A.,** Heart valve replacement with the Ionescu-Shiley pericardial xenograft, *J. Thorac. Cardiovasc. Surg.,* 73, 31, 1977.

165. **Heynan, Y. G., Dureau, G., and Eloy, R.,** Immediate consequences and long term biological consequences of the glycerol-pretreatment of the autologous pericardial valves, *Proceedings of the First World Biomaterials Congress,* Winter, G. D., Ed., Baden, Austria, 1980, 2.9

166. **Kasp-Grochowska, E., Kingston, D., and Glynn, L. E.,** Immunology of bovine heart valves. I. Cross-reaction with the C-polysaccharide of *Streptococcus pyogenes, Ann. Rheum. Dis.,* 31, 282, 1972.

167. **Kasp-Grochowska, E., Kingston, D., and Glynn, L. E.,** Immunology of bovine heart valves. II. Reaction with connective tissue components, *Ann. Rheum. Dis.,* 31, 290, 1972.

Chapter 3

# BIOLOGY, BIOTECHNOLOGY, AND BIOCOMPATIBILITY OF COLLAGEN

## Eric E. Sabelman

### TABLE OF CONTENTS

# I. INTRODUCTION

Collagen is widely known as the principal extracellular structural material of the higher animal body. Since it is immunologically benign, resistant to proteolysis, a natural substrate for cell adhesion, and the major tensile load-bearing component of the musculoskeletal system, collagen is an obvious choice for the manufacture of implantable prostheses. It may even be true that autografts, except those in which vascularity and cell viability are intentionally maintained, derive their utility as frameworks for tissue regeneration from the residual collagenous matrix. If so, it may be said that all grafts are collagen implants, suffering to a greater or lesser degree from contamination.

It is the object of this chapter to examine collagen from which such "contamination" is intentionally removed or inactivated. This object is approached by first reviewing the biochemistry of collagen, its polymerization after *de novo* synthesis, its relationship to other matrix macromolecules, and its turnover and enzymatic degradation. Prospective designers of collagen-based devices are encouraged to take note of the fact that collagen is normally present as a member of a composite structure, and efforts to create tissue-like prostheses composed solely of collagen may be futile. Despite its relatively high intrinsic tensile strength and modulus, and thus its appropriateness as a primary load-bearing member, collagen in normal skin, artery, and tendon is often unstressed under physiological conditions. It seems that weaker materials in the composite structure and electrohydraulic forces are adequate, and the superior properties of collagen are reserved for extreme demands.

The technology of processing is the next aspect of collagen as a biomaterial to be reviewed. Although intact tissues, from which noncollagenous impurities have been removed, are mentioned, the major focus is on the methodology of creating, reprecipitating, and modifying collagen fiber dispersions and monomer (formerly "tropocollagen") solutions. Selection of processing methods affects the end product in not always easily foreseeable ways; compromises have to be made between, for example, ease of cross-linking and possibility of leaching of toxic residuals, or between assurance of sterility and degradation of the collagen polymer.

Representative areas of current research and clinical interest are next surveyed, including cardiovascular prostheses — perhaps the most demanding environment in which collagen is expected to function — bone, tooth and tendon implants, wound dressings (the in vivo application most akin to cell culture), and injectable suspensions, which have the largest treated patient population of any collagen product. In vitro applications include supports for immobilized enzymes with both industrial and medical uses, and substrates for cell culture, nominally an area of pure research into cell-matrix interaction, but providing data applicable to the design of in vivo devices.

Finally, the question of the feedback mechanism between the extracellular matrix and the cells that create it is explicitly addressed. This question may seem far removed from the surgical suite in which the collagen-based implant is sutured to replace a damaged tissue, but the function of that prosthesis may be compromised by lack of understanding of the mode of its recognition by the body and the effect of its geometry at the cellular scale. As a subset of this problem, the property of piezoelectricity is reintroduced; whether or not it is a means of transmitting information on mechanical stress to the cell under normal circumstances, perhaps it could be used to enhance or regulate some other function of a manufactured device such as an immobilized enzyme.

Collagen is a highly versatile material, capable of being formed both into cross-linked compacted solids that are impermeable to cells and enzymatic attack, and thus of long persistence in vivo, and also into lattice-like gels along which cells glide, using pseudopodia to thrust between the collagen strands or anchoring by membrane-bound receptors not unlike those that transmit hormonal signals to the nucleus. Since collagen is molecularly the same in each case, it is tempting to assume that it can possess both kinds of biocompatibility simultaneously, an assumption that is, at best, premature.

## II. BIOCHEMISTRY AND MICROSTRUCTURE

A detailed examination of collagen molecular structure and synthesis is beyond the scope of this chapter; these topics are well covered in classic compendia such as that edited by Ramachandran and Gould,[1] reviews by Eyre,[2] Gay and Miller[3] and Bornstein,[4] and specialized journals. The relevance of collagen biochemistry to biocompatibility lies in its relation to antigenicity, fibrillar reassembly or polymerization, occurrence of post-assembly cross-linking, dehydration or calcification, and the relation of molecular and supramolecular structure to physical properties.

### A. Molecular Types

At least seven types of mammalian collagen have been described.[5] Their common characteristic is a three-stranded helix — each strand a polypeptide rich in glycine. The collagens vary in their nonhelical regions, hydroxylation, glycosylation, and intra- and interhelical cross-linking.[2]

Collagen types IV and V are found in epithelial basement membrane and possess nonhelical regions which impose a mesh-like rather than fibrous ultrastructure.[6] Coupled to type IV collagen are laminin, fibronectin, and heparan sulfate,[7] which interact with a presumptive membrane-bound receptor protein, regulating cell adhesion[8,9] (Figure 1). Components of the basement membrane are resynthesized by migrating epithelial cells;[10] nonetheless, adhesion of these cells is promoted by addition of fibronectin.[11] Type IV has seen only limited use as an in vitro biomaterial.[12,13]

Type III collagen is composed of three identical chains capable of fibrous polymerization, and is present in rapidly growing tissue, particularly juvenile and healing skin. It is a contaminant of preparations of other collagens unless differential salt precipitation is performed to effect separation.[3,13]

Type II collagen also contains identical chains, having more hydroxylysine than other types. It is characteristic of cartilage and the vitreous humor, in which it is part of a multicomponent system, the other major fraction being the glycosaminoglycan (GAG) complex chondroitin sulfate/hyaluronic acid.[14] The spatial interaction of collagen and GAG is a major determinant of cartilage mechanical properties,[15] which is of significance in the persistence of cartilage grafts.[59,190]

Type I collagen is the principal extracellular material of skin, tendon, and bone and tooth matrix; where "collagen" is referred to without prefix, it is understood to mean type I. Its monomer is helix of two $\alpha 1$ chains, each 1055 amino acid residues long, and one, more hydrophilic, $\alpha 2$ subunit. The helical form originates in hydrogen bonding related to the periodic glycine residues, but is stabilized by covalent bonds between lysines; cross-linked chains appear in electrophoresis gels as a "beta" fraction. In its final form, type I has short (totaling 5%) nonhelical telopeptide regions at either end which participate in polymerization by noncovalent binding to sites on adjacent helixes.[16] As synthesized in the form of procollagen, the telopeptides are extended and contain disulfide bridges interlocking the three chains; these are cleaved during secretion by specific peptidases.[4] Hence, denaturation or uncoiling of the mature helix (yielding gelatin) is practically irreversible.

### B. Biophysics of Type I Collagen

Polymerization ordinarily produces fibrils composed of monomers overlapping over 4/5 of their lengths. Since the monomer is 300 nm long, this "D-stagger" array yields a characteristic repeating band pattern at 67-nm intervals.[17] The fibrils are stabilized by interhelical cross-linking catalyzed by lysyl oxidase; other forms of cross-linking, some tissue specific,[18] occur upon aging.[19] Lysyl oxidase can be inhibited by lathyrogens such as $\beta$-aminopropionitrile (BAPN).[20] Collagen from animals fed BAPN is more acid-soluble than

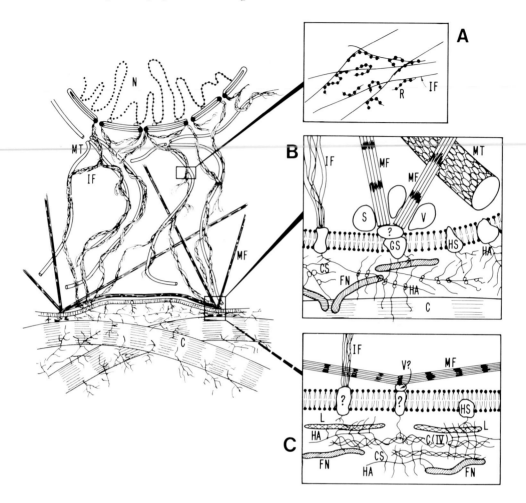

FIGURE 1.    Postulated coupling of cell function to state of the extracellular matrix. N = nucleus; MT = microtubules; IF = intermediate filaments; MF = microfilaments; C = collagen, Type I, II or III. (A) Polyribosome attachment to cytoskeleton. R = ribosomes; (B) nonepithelial cell attachment to collagen. V = vinculin; S = *src*-coded protein kinase; GS = ganglioside attaching fibronectin to membrane; FN = fibronectin; HA = hyaluronic acid; CS = chondroitin sulfate; HS = heparan sulfate; (C) epithelial membrane attachment to matrix. L = laminin; C(IV) = collagen, Type IV. (From Bissell, M. J., Hall, H. G., and Parry, G., *J. Theor. Biol.*, 99, 31, 1982. With permission.)

normal, but the generalized connective tissue disorder mitigates against use of lathyritic animals as sources of collagen.

Since collagen will autocatalytically polymerize given the proper conditions of pH, temerature, and ionic strength, extrapolation to minimal postsecretion cellular influence on fibril assembly in vivo is tempting. However, there is ample evidence that nucleation and axial orientation are initiated and polarized at the time of secretion.[21] Cell-mediated assembly is one side of the feedback loop regulating connective tissue morphology, to be discussed further at the conclusion of this chapter.

Fibrillar packing geometry of dense collagenous tissues has been the subject of some scrutiny. Traub[22] examined interactions between chains of adjacent helices, particularly those regions conserved between species, with the conclusion that hydrophobic interactions are dominant. The same conclusion was drawn by Piez,[17] who discussed supercoiling and alternative configrations of the unit cell of the packed fibril.

The effect of electrostatic, hydrophobic interactions should be emphasized, since the resulting low-entropy hydration sphere surrounding molecular collagen regulates the polymerization rate.[23] Water in bone and tendon collagen is ordered into five regimes, two of which, representing 10% of dry collagen mass, are within the triple helix and contribute to reduction of helix diameter from 1.45 to 1.09 nm upon drying.[24] These observations suggest that dehydration of collagen is irreversible, which may be advantageous, resulting in increased wet strength. On the other hand, it could be deleterious, by unpredictably altering cell responses to the biomaterial.

## C. Calcification

It is at the scale of the fibrillar banding pattern that the participation of collagen in calcification becomes evident. Even collagenous tissues that ordinarily do not calcify if implanted ectopically will precipitate crystals of calcium phosphates at 67-nm intervals,[25] corresponding to the gap between the amino terminal end of one helix and the carboxyl terminus of the next.[26] As is the case with autopolymerization, although calcification can occur spontaneously, it is ordinarily a cell-mediated process, governed by changes in inhibitory glycosaminoglycans[27] and/or promotors, such as osteonectin,[28,29] at the mineralization front in bone, or by membrane-bounded vesicles which act as crystallization nuclei in the cartilaginous epiphyseal plate.[30] Elastin[25] and poly-hydroxyethylmethacrylate[31] are other biopolymers capable of initiating calcification. It is important to distinguish between calcification of preexisting collagen and ossification, the process of *de novo* secretion and subsequent mineralization of collagenous matrix; the latter may be induced by noncollagenous components of implanted demineralized bone or tooth[32] or by migration of osteoprogenitor cells into pores of a calcified implant.[33]

Calcification is one mechanism by which collagen acts as a component in composite structures having highly varied mechanical properties. The assumption that collagen behaves as does the fibrous component in a synthetic composite is often inadequate. The strength of bone is a function of mineral density, not only because of stiffening due to spatial constraint of collagen fiber buckling and innate compressive strength of the mineral phase, hydroxylapatite, but also because of covalent cross-links between collagen and inorganic phosphate.[2,34]

## D. Composite Structure

Tendon and ligament are also examples of collagen-containing composites, despite being composed of nearly "pure" collagen. Skin has a comparable dry weight fraction of collagen (60 to 75% in human skin vs. 75 to 85% in bovine tendon); however, its larger proportion of elastin, glycosaminoglycan content, and quasirandom orientation yield a material having substantially different small-strain properties.[35] Human ligament varies in modulus, ultimate strain, and rupture stress at different locations in the same joint.[36] The same properties of rat tail tendon increase monotonically with age[37,38] (Table 1). During aging, collagen volume increases relative to cell population, GAG, elastin, and water content; tenocyte cell morphology alters and synthetic activity declines, while contractile capacity is unchanged.[41]

Active contraction may be responsible for the zigzag crimping observed in relaxed rat tail tendon.[42] Crimping apparently maintains registry of fascicles as they slip past one another during bending.[43] If rat tail tendon is comparable to ligaments having higher elastin contents (2 vs. 20%), crimp in the tail tendon and spiral microstructure in bovine ligamentum nuchae are maintained by tension in elastin, which effectively shields collagen fibers from tensile loading at low strains (4 to 70% elongation).[44] The interaction of elastin and collagen in determining physical properties is more apparent in vascular radial compliance than in tendon elongation.[45] The conclusion to be drawn from this examination of natural tissues is that collagen in its unalloyed state, produced by either reconstitution or extraction, cannot be expected to behave as an intact tissue.

## E. Degradation

Because of its helical structure and packing, collagen is not susceptible to enzymatic

## Table 1
## MECHANICAL PROPERTIES OF COLLAGENOUS MATERIALS

| Material | | Tensile yield (MPa) | Elastic modulus (MPa) | Ref. |
|---|---|---|---|---|
| Human skin, fresh | | 0.2 | 1.0 | 39 |
| Human artery, fresh | | 0.4 | 348 | |
| Rat tail tendon | | | | |
| 1 month | | 20 | 337 | 37 |
| 12 month | | 67 | 460 | |
| 18 month | | 73 | 489 | |
| Rhesus medial collateral ligament | | 73 | 463 | 36 |
| Anterior cruciate | | 66 | 186 | |
| Human ligament and tendon (50 cm/min) | | | | |
| Medial collateral | | 8 | 327 | 40 |
| Flexor hallucis longus | | 20 | 316 | |
| Extensor hallucis longus | | 27 | 387 | |
| Peroneus brevis | | 21 | 668 | |
| Tibialis posterior | | 16 | 920 | |
| Dried-rehydrated collagen tape | | 2.8—127 | 24—578 | 39 |
| Xenograft | Strain rate | | | |
| "LR" rate | 5 cm/min | 15 | 827 | 40 |
| | 50 cm/min | 17 | 781 | |
| "CLR" rate | 5 cm/min | 33 | 2924 | |
| | 50 cm/min | 27 | 2363 | |
| "GR" rate | 5 cm/min | 43 | 2658 | |
| | 50 cm/min | 28 | 1531 | |

digestion except by type-specific collagenases, which bind to recognition sites on the three polypeptide chains.[46] Collagenase is present in undisturbed tissues in a latent form, and is activated by proteases, as demonstrated by prolonged collagen half-life in granulation tissue on administration of plasmin and trypsin inhibitors.[47] Fibroblastic collagenase synthesis is controlled by an interleukin-like factor released by macrophages[46] and stimulated by factors such as heparin.[48] Breakdown products of collagen itself can stimulate collagenase production.[49] Collagenase activity is inhibited by alpha-macroglobulin, platelet factor, and tissue-specific factors,[48] as well as by cross-linking of the substrate.

Collagen degradation is a normal aspect of wound healing, preceding new collagen synthesis by 3 to 5 days.[50] Macrophage immigration into a wound is apparently stimulated by diffusion of collagen breakdown products from intrinsic and bacterial collagenolytic activity.[51] Collagen is normally replaced in bone, tendon, and skin at the same rate it is degraded; alterations in metabolism or function can greatly accelerate turnover, often resulting in excessive resorption, as in renal osteodystrophy or denervation atrophy.[52] In contemplating the use of collagen as an implanted biomaterial, one must bear in mind that it will be subject to the same turnover as normal tissue, plus whatever enhancement of degradation occurs as a result of geometric or immunologic feedback and healing of the implantation incision.

## F. Immunology

It is generally agreed that collagen is only mildly immunoreactive, due in part to masking of potential antigenic determinants by the helical structure.[53] The major repository of accessible antibody binding sites is the amino-terminal telopeptide,[54] as would be expected from recent findings on the enhancement of antibody recognition by atomic mobility.[246] A review by Furthmayr and Timpl[54] points out the highly species-specific ability to generate antibodies to helical as compared to terminal regions. The rabbit is particularly sensitive to

## Table 2
### COMPARATIVE PROCESSES FOR INTACT COLLAGENOUS TISSUES

| Material | Treatment | Result | | Ref. |
|---|---|---|---|---|
| Cat costal cartilage | Fresh autograft | Continued growth | | 59 |
| Cat costal cartilage | Freeze-thawed autograft | Resorption only at cut end | | |
| Cat costal cartilage | 3-4 Mrad irradiated allograft | 25—50% resorption at 14 months | | |
| | | **Strain @ 1 N/mm²** | **Suture Pull-out** | |
| Human saphenous vein | 100% glycerol | 54% | 8.8 N/mm | 58 |
| Human umbilical vein | Fresh | 35% | 6.7 | |
| Human umbilical vein | 1% glutaraldehyde | 33% | 4.7 | |
| Bovine carotid artery | 1. Buffered ficin digest<br>2. Adipyldichloride cross-link<br>3. Glutaraldehyde cross-link<br>4. Ethylene oxide sterilize<br>5. Store in saline | 20% ultimate strain;<br>39% occlusion after 6 week implant<br>vs. 77% for Dacron | | 60 |
| Rat infrarenal aorta | 1. Insert Silastic mandrel<br>2. Trypsin digest<br>3. Add GAGs, 50—100 µg/artery<br>4. Aldehyde vapor cross-link<br>5. 0.5% lysine HCl quench | Wall thickness 40% of fresh | | 61 |
| Rat dorsal skin | 1. Trypsin digest, 2 mg/mℓ, pH 9 +<br>NaN₃ 0.5 mg/mℓ, 28 days @ 15°C<br>2. Glutaraldehyde cross-link, 0.01%,<br>16 hr @ 15°C<br>3. Store in TRIS-buffered saline, pH<br>7.2 + NaN₃<br>4. Rinse 2x in Medium-199 | Allogeneic implant 60—90% intact at<br>8½ weeks vs. 50—60% allograft | | 62 |

xenogeneic amino acid sequences in nonhelical regions, only 20% of determinants being on the helix.

Antigenicity can be enhanced by incorporation of collagen in Freund's adjuvant, covalently bound polytyrosine, methylation, succinylation, or similar alteration of the exposed helical chains.[54] The determinant sites of xenogeneic collagen can be further inhibited by cross-linking,[55] perhaps by preventing diffusion of cleaved chain fragments. Antigens on cleavage fragments probably interact with T cells at a wound or implant site prior to antibody production;[56] the clinical correlate is a localized inflammation rather than systemic response unless the recipient is presensitized. The relative inertness of pure collagen may be contrasted with clinical evidence of immunoreactivity of multicomponent allografts. Lee et al.[57] reported that 91% of patients receiving osteochondral grafts had detectable levels of cytotoxic antibodies beginning at 3 weeks postsurgery.

## III. EXTRACTION AND PURIFICATION

### A. Intact Tissues

If autografts and frozen or freeze-dried allografts are excluded, the simplest collagenous implants from a process standpoint are intact tissues from which noncollagenous "contaminants" have been extracted or inactivated. Comparison of different published processes is difficult, since seldom are mechanical, immunological, and implantation tests performed on identically processed material.

Table 2 gives examples of various processes intended to result in functional implants. Bennett and Drury[58] reported on effects of single-step treatments of vascular grafts on mechanical properties; glycerol preservation yielded a material having 54% lower average tensile modulus and 31% higher suture pull-out strength, while glutaraldehyde preservation resulted in unchanged modulus but reduced strength by 30% compared to fresh veins. Gerolanos et al.[60] described a commercial product, "Solcograft-P", in which noncollagenous components were extracted by ficin digestion, followed by two-step cross-linking, the first to establish long interfibrillar cross-links and the second to bind antigenic sites. The process used by Gerolanos was apparently derived from experiments by Rosenberg et al.,[63] who compared unbuffered and buffered formalin, polyacrolein, and dialdehyde starch cross-linking of ficin-digested bovine carotid artery, with the latter two methods yielding higher heat-shrinkage temperatures. Loissance et al.[61] used trypsin rather than ficin, as did Oliver and co-workers.[62] Trypsin is more specific, cleaving at basic amino acids, and leaves some elastin intact.[64] Loissance, aware that loss of glycosaminoglycans would affect mechanical properties, attempted to reintroduce GAGs prior to cross-linking.

## B. Dispersion and Dissolution

One class of processes yielding collagen in fluid form suitable for reconstruction into a solid is dispersion, which is taken to mean comminuition and swelling, possibly accompanied by partial solubilization (Table 3). A second category of fluidized collagen continues disaggregation to the point of depolymerization; any undissolved material is separated from the monomer solution by filtration or centrifugation.

The source material determines the inital steps in processing; hide contains more fat than tendon, which is removed by flotation and skimming, or by alkaline or detergent saponification.[66] Noncollagen proteins and cell debris are extracted by enzymatic digestion (for intact tissues) by prolonged acid treatment, or by neutral salt extraction.[76] The acid dissolution stage alone suffices to dissolve newly synthesized rat tail tendon.[69]

If a solution, rather than dispersion, of mature collagen is desired, it is necessary to enzymatically cleave the telopeptides and thus release the monomer from interhelical cross-links.[78] Pepsin[13] and trypsin (e.g. Becker et al.[66]) are frequently used; pepsin leaves a fragment of telopeptide which may facilitate repolymerization. Proctase and pronase are less specific, but have been successfully employed, particularly by Miyata and colleagues.[79]

Chvapil[80] reviewed properties of dispersions and solutions, and pointed out the advantages of dispersion and slurries in higher solids content for equivalent viscosity (1 vs 0.1%, respectively, are pumpable fluids), potential for reprecipitation from acid by dehydration, and higher strength due to retention of cross-links and telopeptides. For implantation purposes, monomer solutions have advantages in potential for filter sterilization, more exhaustive purification, and excision of telopeptides and their antigenic sites. In designing a process for an implantable device, these features must be weighed with regard to their impact on the intended function and permanence of the device.

Complete removal of glycosaminoglycans from dispersed collagen is not possible without dissociative reagents such as guanidine-HCL or 8 $M$ urea. Such exhaustive extraction is seldom considered necessary except in the case of bone collagen. The reprecipitation protocol followed by Danielsen[75] (Table 3) yields a highly purified monomer solution, having only 0.04% residual uronic acid (indicative of unextracted GAGs). Other methods rely on the varying isoelectric points of different collagens, including the buffered salt precipitation of Becker et al.[66] and the proprietary technique employed by Collagen Corporation.

Protease inhibitors such as PMSF[75] or aprotinine[66] have recently been included, presumably to reduce autolysis of telopeptides and endogenous collagenase activity; these are superfluous if later protease solubilization is part of the protocol. Bactericidal agents[70] may be indicated for long salt extractions, but acids at the concentrations employed are often sufficiently bacteriostatic and obviate the problem of residuals in the end product. Enzymatic solubilization introduces the similar problem of enzyme carryover; enzymes are inactivated by

## Table 3
# COMPARATIVE PROCESSES FOR DISPERSED AND SOLUBILIZED COLLAGEN

| Material | | Process | Ref. |
|---|---|---|---|
| Cowhide | 1. | Soak, rinse in water; grind | 65 |
| | 2. | 1% NaOH + 0.3% $H_2O_2$, 24 hr | |
| | 3. | (Optional) 0.01% *B. subtilis* enzyme + 0.04% urea + 0.01% $(NH_4)_3SO_4$, 5 days | |
| | 4. | HCl, 10% acid/hide w:w, 6 hr; wash | |
| Human umbilical cord | 1. | Cut to 1 cm | 66 |
| | 2. | Tris-HCl, pH 7.4 + 0.5 *M* NaCl + 0.01 *M* EDTA +/− aprotinine (protease inhibitor), 24 hr @ 4°C | |
| | 3. | Precipitate: 0.5 *M* acetic acid, 24 hr | |
| | 4. | Supernate: 0.9 *M* NaCl, 2 hr | |
| | 5. | Precipitate: Tris-HCl, pH 7.4 + 1 *M* NaCl | |
| | 6. | Dialyze 48 hr against above buffer; centrifuge; retain supernate | |
| Bovine tendon | 1. | Slice to 275 μm | 67 |
| | 2. | 0.1% ficin + 0.001 *M* $Na_4$-EDTA, 12 hr @ 20°C | |
| | 3. | $H_2O_2$ (inactivate ficin), 1/2 hr | |
| | 4. | 0.2% cyanoacetic acid +/− 50% methanol, 3 hr < 25°C; homogenize | |
| Bovine tendon | 1. | Slice to 275 μm | 68 |
| | 2. | 3% malt diastase, pH 7, 15—20 hr @ 37°C | |
| | 3. | 0.04% EDTA, 2 hr @ 37°C | |
| | 4. | 0.35% perfluorobutyric acid + 50% methanol, 3 hr; homogenize | |
| Rat tail tendon | 1. | Separate fascicles; mince | 69 |
| | 2. | 0.1% acetic acid, 48 hr @ 4°C; centrifuge | |
| | 3. | Store supernate @ −70°C | |
| Calf skin | 1. | 5% $NH_4Cl$ + 1% NaCl + 0.01% merthiolate, 3 days @ 20°C | 70 |
| | 2. | Grind; dialyze against water @ 4°C | |
| | 3. | 0.5 *M* acetic acid, 24 hr @ 4°C; store filtrate | |
| Rat tail tendon | 1. | Whole tendons | 71 |
| | 2. | 0.5 *M* $NaH_2PO_4$, 3 days @ 4°C; rinse | |
| | 3. | Soak in water, 12 hr @ 4°C; store supernatant, pH 4.1 | |
| Bovine tendon | 1. | Slice tendons | 72 |
| | 2. | 2.5—5% acetic acid, 4 hr | |
| | 3. | Dilute to 0.1% collagen (1:8 in acetic acid); filter | |
| | 4. | Neutralize with 1 *M* $NH_4OH$; wash | |
| | 5. | Disperse precipitate in 6 g/ℓ malonic acid, to 0.5% collagen | |
| Rat tail tendon | 1. | 0.5 *N* acetic acid, 2 days @ 4°C; centrifuge | 73 |
| | 2. | Supernate: 20% NaCl, 1:1 v:v; centrifuge | |
| | 3. | Precipitate: redissolve in 0.5 *N* acetic acid | |
| | 4. | Dialyze against water, 2 days, to $5 \times 10^{-5}$ *N* acetic acid; centrifuge; store supernate | |
| Rat tail tendon | 1. | Wash 1% NaCl, all steps @ 4°C | 74 |
| | 2. | 0.5 *N* acetic acid, 12 hr; centrifuge | |
| | 3. | Supernate: add dry NaCl to 20% w:v; centrifuge | |
| | 4. | Precipitate: 0.5 *N* acetic acid; dialyze against same | |
| | 5. | 30% NaCl + 0.5 *N* acetic acid, 1:5 v:v, 2 hr | |
| | 6. | Precipitate: wash 0.02 *M* $Na_2HPO_4$ | |
| | 7. | Redissolve in 0.5 *N* acetic acid; dialyze as above, 2 days | |
| | 8. | Store supernate, 1—4 mg/mℓ | |
| | 9. | Adjust to 2 mg/mℓ; add pepsin 1:10 w:w collagen, 2 × 24 hr | |
| | 10. | Dialyze against 0.02 *M* $Na_2HPO_4$; centrifuge | |

**Table 3 (continued)**
## COMPARATIVE PROCESSES FOR DISPERSED AND SOLUBILIZED COLLAGEN

| Material | | Process | Ref. |
|---|---|---|---|
| | 11. | Precipitate: redissolve in 0.5 $M$ acetic acid | |
| | 12. | Dialyze against 0.01 $M$ acetic acid; lyophilize | |
| Lathyritic rat skin | 1. | Steps 1, 3—8 as above | 74 |
| | 2. | 1 $N$ NaCl + 0.05 $N$ Tris-HCl, pH 7.5, 2 × 4—5 hr | |
| Rat dorsal skin | 1. | Mince; wash in 1% NaCl | 75 |
| | 2. | Homogenize in 1 $M$ NaCl + 0.05 $M$ phosphate, pH 7.5 + 0.02 $M$ EDTA + 0.01 $M$ PMSF; stir 12 hr @ 4°C | |
| | 3. | Precipitate: 0.5 $M$ acetic acid, 5 days @ 4°C | |
| | 4. | Supernate: 5% NaCl | |
| | 5. | Redissolve precipitate in 0.5 $M$ acetic acid | |
| | 6. | Dialyze against 0.02 $M$ phosphate, pH 7.5 | |
| | 7. | Redissolve precipitate in 0.6 $M$ NaC$_2$H$_3$O$_2$, pH 4.8 | |
| | 8. | Centrifuge; retain supernate | |
| | 9. | Dialyze against deionized water; lyophilize precipitate | |
| Rat tail and skin | 1. | 1:100 in 0.45 $M$ NaCl, 24 hr | 13 |
| | 2. | 1:100 in dilute acetic acid, 48 hr @ 4°C | |
| | 3. | Supernate: reprecipitate in 0.02 $M$ NaHPO$_4$ | |
| | 4. | 0.1—1 mg/m$\ell$ pepsin (1:100 enzyme/substrate), pH 2, 6 hr @ 4°C | |
| | 5. | Dialyze against Tris buffer, pH 7.4, 24 hr<br>a. + 0.45 $M$ NaCl: Type I precipitates<br>b. + 1.7 $M$ NaCl: Type III precipitates<br>c. Without salt: residual Type I precipitates | |
| | 6. | Dialyze against 0.02 $M$ NaH$_2$PO$_4$: Type IV precipitates | |

*Note:* Repetitions of processing steps, water rinsing, and mechanical operations such as stirring not listed; EDTA = ethylene diamine tetra-acetic acid (chelating agent); PMSF = phenylmethanesulfonylfluoride (protease inhibitor).

oxidation[67] or pH change, and remain in the supernate upon subsequent precipitation/dissolution purification.

## C. Precipitation

Precipitation into a final physical form is the usual endpoint of collagen processing. Exceptions exist in the use of dispersions as stabilizers for pharmaceutical[81] and cosmetic emulsions, and in the polymerzation and subsequent homogenization of monomer solutions for use as injectable suspensions.[77]

Precipitated dispersions have seen their greatest application in sutures,[83,84] membranes,[72,82] and sponges.[65,85] Collagen is most frequently deposited from solution as a coating[86] or gel. A survey of precipitation methodology is given in Table 4.

Dispersions are exclusively precipitated by dehydration, either air drying[72,83] or osmotic dehydration in concentrated salt solutions[82] or water-miscible organic solvents.[84] Fabrication of sponge involves dehydration of a dispersion after freezing;[65,85] the freezing rate determines ice crystal size and thus pore diameter.

Acid solutions of monomer may be air-dried, but lack well-formed fibers[78,86] (Figure 2A and C). Postdrying neutralization is by dilution in tissue fluid, culture medium, or water rinses. Historically, neutralization to initiate polymerization has been by dialysis against buffered saline.[70] More recently, it has become common to add neutralization buffer directly to collagen solution, either as part of a culture medium.[73,86] (Figures 2B and D and

## Table 4
## COMPARATIVE PROCESSES FOR COLLAGEN PRECIPITATION FROM DISPERSIONS AND SOLUTIONS

| Source tissue final form | | Process | Ref. |
|---|---|---|---|
| Bovine tendon | 1. | 20% "solubilized" + 80% "fiber", acid dispersion | 82 |
| | 2. | Sat. NaCl until coagulated | |
| | 3. | Neutralize with NaOH; wash; air dry | |
| Membrane, tube | 4. | UV cross-link | |
| Bovine tendon | 1. | 0.5% collagen in malonic acid | 72 |
| | 2. | Add to fabric-lined centrifuge | |
| | 3. | $NH_4OH$, 1 1/2—2 hr; vary RPM | |
| | 4. | (Optional) 2% formaldehyde, 2 × 1/2 hr | |
| 75 μm thick membrane | 5. | Air dry in centrifuge | |
| Rat tail tendon | 1. | 0.7 mg/mℓ collagen in acetic acid, 50 μℓ/35-mm dish | 86 |
| Coated petri dish | 2. | $NH_3$ vapor, 1/2 hr | |
| Coated petri dish | 2. | Riboflavin-5'-phosphate, 0.01% + UV, 1/2 hr | |
| Coated petri dish | 2. | Air dry | |
| Gel | 2. | 10 μℓ/dish 6% NaCl, 4 hr @ 37°C | |
| "Vitrogen" (cowhide) | 1. | 2.5 mg/mℓ in HCl, 50 μℓ/9-mm well | |
| Thick gel | 2. | L15 medium, 1:1, neutralize with NaOH, 4 hr @ 37°C | |
| Rat tail tendon | 1. | 0.2 mg/mℓ solution in 0.005 *M* acetic acid @ 5°C | 87 |
| | 2. | NaCl, 16 g/ℓ + Tris-HCl, 14 g/ℓ + $Na_2HPO_4 \cdot 7H_2O$, 16 g/ℓ + NaOH to pH 7.45—7.5, 1:1 v:v, ionic strength 0.225 | |
| Gel | 3. | Warm to 26°C | |
| Calf skin | 1. | 0.1—0.37% solution in acetic acid | 70 |
| | 2. | Dialyze against phosphate buffer, pH 7.6, 24—48 hr @ 3°C | |
| Gel | 3. | Warm to 37°C in 1/2 hr | |
| Rat tail tendon | 1. | 0.05% solution in acetic acid + merthiolate | 71 |
| | 2. | Evaporate from dialysis tubing to 0.05—0.4% | |
| Gel | 3. | Phosphate buffer, pH 7.5, ionic strength 0.5 | |
| Rat skin | 1. | 0.005 *M* acetic acid solution, 1.0—1.4 mg/mℓ | 75 |
| | 2. | 0.2 *M* NaCl + 0.11 *M* Tris-HCl, pH 7.4, 1:1 v:v, @ 6°C | |
| | 3. | Warm to 30°C @ 4 deg/hr | |
| | 4. | Dialyze against water, 120—140 hr @ 37°C | |
| Membrane | 5. | Air dry, 4—6 days @ 37°C; age 11—104 days | |
| Cowhide | 1. | HCl suspension; washed | 65 |
| | 2. | Long fibers: 10% NaCl, w:w, 2 hr; centrifuge | |
| | 3. | Short fibers: homogenize washed suspension | |
| | 4. | Mix long and short fibers 2:3; dilute to 2.2% collagen, pH 9 | |
| | 5. | 4% Triton X-100 (anionic detergent) + 2% hexamethylene diisocyanate | |
| | 6. | Freeze, 12 hr @ −25°C | |
| Sponge | 7. | Lyophilize; split to 2-mm thick | |
| Rat tail tendon | 1. | 50 μℓ acetic acid solution per 60-mm petri | 69 |
| | 2. | 10 μℓ 6% NaCl, 1 hr @ 37°C | |
| Gel coating | 3. | Rinse: water, isotonic saline | |
| Bovine tendon | 1. | 1% suspension in methanol-cyanoacetic acid | 83 |
| | 2. | Extrude through 0.1 - mm slit at 36 cm/min | |
| | 3. | Air dry @ 75°F, RH 55% | |
| | 4. | Repeat for multiple laminae | |
| Oriented sheet | 5. | Methanol wash; redry | |
| Bovine tendon | 1. | 1% suspension in methanol-cyanoacetic acid | 84 |
| | 2. | Age 31 hr @ 25°C, then 16 hr @ 5°C | |
| | 3. | Extrude into circulating acetone + $NH_4OH$, 138 mg/ℓ | |
| Multifilament suture, 5/0 | | or methyl-ethyl-ketone + $NH_4OH$, 43 mg/ℓ | |

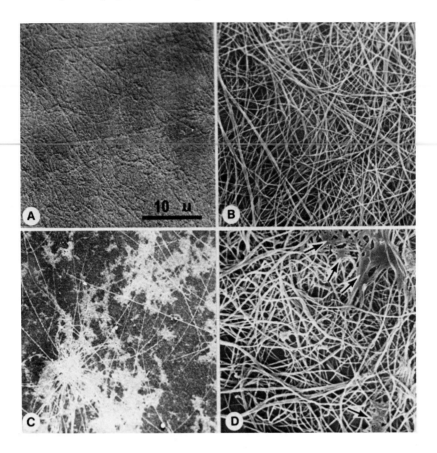

FIGURE 2.    Equal magnification scanning electron micrographs of representative regions of
(A and C) air-dried thin collagen coating; (B and D) thick NaOH-neutralized physiologic ionic
strength (mixed 1:1 with medium L-15, 4 hr @ 37°C) collagen gel. Collagen source was either
"Vitrogen 100" atelopeptide bovine skin (Collagen Corp., Palo Alto, Calif.), 2.5 mg/mℓ (A
and B) on rat tail tendon, 0.74 mg/mℓ (C and D). Arrows in D show nonfibrous protein in rat
tail solution. (From Iversen, P. L., Partlow, L. M., Stensaas, L. J., and Moatamed, F., *In Vitro*,
17, 540, 1981. With permission.)

Figure 3) or as a 10:1 or 2:1 concentrate.[75,87] The resulting gel can be dried to form a
membrane that is 95% soluble in 80°C water after 2 hr.[88] The three-dimensional fiber mesh
collapses upon drying; incubation in 0.005 $M$ acetic acid for 30 min at 20°C is incapable
of restoring the structure.[89]

The early work of Bianchi[90] on effects of various salts on denaturation temperature, and
the studies of Piez and co-workers[91] have elucidated the roles of temperature and ionic
strength in reassembly of the native fibril. A temperature-dependent initiation step is followed
by non-temperature-dependent linear growth of filaments, then by temperature-sensitive
diametral growth or aggregation of thin (2 to 8 nm diameter) fibrils.[91] The first two factors
introduce a time lag ranging from 10 min to 10 hr before a turbidity plateau indicative of
diameters large enough (60 to 200 nm) for significant light scattering is reached.

The kinetics of fiber nucleation and growth are in accordance with the Avrami equation
governing polymer crystal formation.[92] Collagen differs from simple polymers in the oc-
curence of time- and temperature-dependent aging, in which spontaneous mechanisms of
entanglement, syneresis, and aldimide cross-linking inhibit reversibility.[93] Other factors are
concentration dependence, which results from a larger number of nucleation sites but restricted radial growth of fibrils at high concentration,[92] and ionic strength dependence, higher

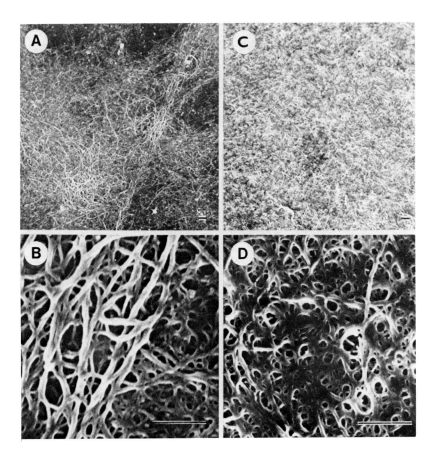

FIGURE 3. Scanning electron micrograph of representative regions of collagen gel. (A and B) Ammonia neutralized, 30 min.; (C and D) salt precipitation (mixed 5:1 with 6% NaCl, 4 hr @ 37°C). Scale bar = μm. (From Iversen, P. L., Partlow, L. M., Stensaas, L. J., and Moatamed, F., *In Vitro*, 17, 540, 1981. With permission.)

salt concentration slowing the onset of turbidity. There is evidence that protease-digested atelopeptide collagen is more sensitive to precipitation conditions than the intact molecule.[91] These factors are relevant to fabrication of collagen-based products, in that fibril diameter and other properties can be selected, to a degree, by control of precipitation conditions.

Two other methods of precipitation deserve mention. Masurovsky and Peterson[94] originated a technique for photopolymerization-coating rat tail tendon solutions using riboflavin-5'-phosphate as a catalyst and exposure to fluorescent light as an energy source. The fibers formed were thin and lacked cross-banding. Iversen and co-workers[86] tested the method using ultraviolet light and found that nodules rather than fibers were formed.

The second method is electrodeposition, which may result from local pH elevation at the cathode, combined with dielectric or electrophoretic alignment of monomer. Marino et al.[95] used 10-msec pulses at 1000 V to deposit calf skin collagen from low-ionic-strength citrate solution on a carbon electrode, resulting mainly in thin, unbanded filaments. Suzuki et al.[76] also described work on electrodeposited collagen-enzyme composite films.

Collagen fibers in gel or dispersion are randomly oriented, having isotropic bulk mechanical properties. The high axial tensile strength of collagen can be taken advantage of by orientation of the fibril bundles. Lieberman and Oneson[72] used fluid shear induced by changing the rotational speed of a centrifuge during precipitation of a dispersion. Frictional shear during extrusion, followed by stretching, was used by Hanyen et al.[83] Stenzel et al.[78]

and others have extruded hollow tubes through counter-rotating nozzles to impart helical fiber orientation. Hughes and colleagues[39] have developed a proprietary process for orienting neutralized monomer in solution during warming from 4 to 20 to 40°C; fiber anisotropy was retained after dehydration to the form of a flat ribbon. It is possible that shear-induced order during polymerization reduces the energy requirement for nucleation and growth, in addition to aligning already formed fibers.[96]

## D. Cross-Linking

Artificially cross-linking a collagen product substantially increases its strength, stiffness, and persistence in vivo. The oldest chemical methods evolved from leather tanning, and persist in the use of chromic suture.[84] Formaldehyde is occasionally used, but results in a relatively brittle and unstable product compared to dialdehydes.[62,78] Glutaraldehyde (0.01 to 0.4%, 1 to 72 hr; see Table 2) yields a satisfactory result from mechanical[98] and persistence[97,99] standpoints. Exhaustive rinsing or dimedone (0.2%, 24 hr)[98] removes unreacted glutaraldehyde. Free aldehydes can also be bound or quenched with an excess of glycine or lysine.[61]

Glutaraldehyde has the property of autopolymerization, becoming resistant to extraction. Birefringent nodules of polyglutaraldehyde may remain in an implanted product despite extensive washing.[100] Autopolymerization is not entirely irreversible, particularly under hydrolytic conditions or low pH. Free aldehyde (6 $\mu$g/g collagen) is extractable in buffered saline (pH 7, 25°, 24 hr), increasing an order of magnitude at pH 4.5.[101] The possibility exists that the prolonged persistence of glutaraldehyde cross-linked collagen in vivo is due as much to toxic inhibition of collagenolysis as to stabilization of fibrillar structure or masking of antigenic sites. Nonetheless, clinically satisfactory results are claimed with glutaraldehyde-preserved intact tissues that are simply rinsed prior to implantation.[102] Where slow release of aldehyde must be avoided, an alternative method of cross-linking should be considered; one possibility is di-isocyanate.[80]

Nonchemical methods of cross-linking exist, depending mainly on radiation-induced free radical formation. Ultraviolet cross-linking was employed by Stenzel et al.[82] both to increase strength of collagen dialysis membrane and to control effective pore diameter; average pore size was 33.6 nm without UV and 25.9 nm after cross-linking. Shimizu and colleagues[103] experimented with UV (1 to 6 hr, 100 W lamp at 10 cm), gamma radiation (1 to 5 Mrad) and glutaraldehyde (saturated vapor or 0.1 to 3%, 3 to 72 hr) for covalently binding collagen to plastic. Growth of human lung fibroblasts in vitro was maximum on untreated and UV-irradiated collagen films and nil on 0.1% glutaraldehyde cross-linked films. Peel strength of films applied to rabbit skin wounds after 2 months was greatest for 3 Mrad gamma irradiated material (80 g/cm), constant at 55 g/cm for all ranges of UV irradiation, and rapidly declined from 45 g/cm to zero after more than 24 hr glutaraldehyde treatment.

## E. Sterilization

The same chemical reactions and radiation-induced radicals that are responsible for cross-linking may also serve for sterilization. In current practice, these are[104]

1.    UV light is absorbed by bases of nucleic acids and aromatic amino acids, and is commonly used for sterilization of collagen-coated petri dishes.[105]
2.    Gamma irradiation from ${}^{60}$Co sources at energies of 1.25 MeV is suitable for commercial-scale applications in the 1.0 to 2.5 Mrad range.
3.    Electron beam radiation is less penetrating than gamma rays, has higher specific energy (12 MeV), and is delivered in a shorter period of time.[106] Reduced exposure time for equal dose apparently results in lower molecular damage to polymers, but entails a 10 to 12°C temperature rise.

4.  Oxidants such as iodides[107] or peroxides[108] react with sulfhydryl and disulfide moieties, in which atelopeptide collagen is depleted.
5.  Alkylating agents, including aldehydes and epoxides, react with amine, carboxyl, hydroxyl, and sulfhydryl groups. Residuals and toxic reaction products are disadvantages, in addition to change in mechanical properties. Even adsorbed on synthetic polymers, glutaraldehyde residuals from a 2% buffered sterilizing solution can be as high as 5 ppm after five rinses.[109] Ethylene oxide gas sterilization similarly requires exhaustive vacuum extraction and/or aeration.[110]
6.  Filter sterilization through 0.22- or 0.45-μm membrane filters is possible only with dilute collagen solutions and entails careful prefiltration to avoid clogging and aseptic handling during subsequent processing, but can be accomplished without alteration of collagen properties.

Chvapil et al.[99] compared sterilization by propylene oxide (4%, 48 hr) with $^{60}$Co irradiation (1.5 Mrad) on lightly or heavily glutaraldehyde cross-linked collagen, with propylene oxide resulting in superior in vivo persistence of lightly tanned material. Reduced persistence after radiation sterilization reflects collagen α-chain breakage and denaturation upon hydration. In Schnell and co-workers'[111] comparison of ethylene oxide and gamma and electron beam sterilization of gut sutures, protein extractable in 0.9% saline increased 4-fold with ethylene oxide or 2.5 Mrad irradiation, reaching 6 to 8% at 10 Mrad, with electron beams tending to cause more damage at higher doses.

In experiments at Collagen Corporation, LiCl-extracted bovine demineralized bone powder was sterilized by 1.0 Mrad gamma irradiation. Extractable carbohydrates and lipids increased four-fold (to 0.01 and 0.2%, respectively), while saline-soluble protein increased from 0.04 to 1.2%. Cross-linked aggregates increased as intact collagen chains decreased (Figure 4).[112] Nonetheless, irradiated collagen appears to perform adequately some in vivo situations.[59,113]

## IV. PRESENT APPLICATIONS

Collagen has a long history of experimental use, as well as a few applications for which clinical utility is well established. Stenzel et al.[78] and Chvapil et al.[55,80] thoroughly reviewed progress through the early 1970s, including European work only lightly touched on here. Other sources[114] cited isolated cases of clinical trial of collagen biomaterials; quite often the result was poorly described and the technique did not remain in the clinical armamentarium.

Rather than repeat the earlier reviews, the following survey consists of selected in vivo and in vitro applications of purified or reconstituted collagen which illustrate how biochemical and physical properties interact with biological systems. These examples are drawn from the broad categories of reconstructive surgery, musculoskeletal and cardiovascular prosthetics, and substrates for cell culture and enzyme immobilization.

A collagen-based implant is subject to the general constraints imposed on any foreign material, in addition to the specific advantages and limitations deriving from its composition. The physical form of a biomaterial may profoundly influence the responses of host tissue; results of implantation tests depend on implantation site, species, animal age, implant size, surface area, and roughness.[115] It is necessary to select test systems appropriate to the end use of the material, and to measure parameters which will enhance function if optimized.

Surface texture is an aspect known to influence cellular response to alloplastic implants,[116] but seldom considered as being responsible for success of biologically derived implants. Geometric difference from the host tissue bed is probably the major factor stimulating foreign body response. For example, particles of calcium phosphate ceramic, otherwise identical, remain surrounded by foreign-body giant cells in an intramuscular site if of a sharp-edged angular profile, while inciting only fibrosis at 6 months if rounded.[117] Matlaga et al.[118]

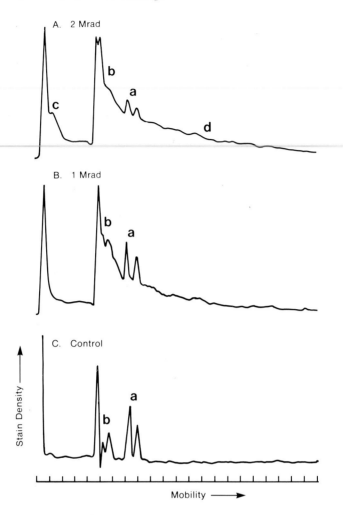

FIGURE 4.  Sodium dodecyl sulfate 15% polyacrylamide gel electrophoresis of gamma-irradiated demineralized bovine bone powder. (A) 2 Mrad dose; (B) 1 Mrad; (C) control. a = intact alpha chains; b = beta (cross-linked) chains; c = highly cross-linked aggregate in exclusion volume; d = degraded peptides. (From Sabelman, E. E., Armstrong, R., and Seyedin, S., *Proc. 9th Meet. Soc. Biomater.*, 6, 22, 1983. With permission.)

implanted intramuscularly in rats extruded medical-grade plastics of circular, pentagonal, and triangular cross-section. At 14 days, macrophage population and lysosomal enzyme content was greatest adjacent to corners and maximal surrounding triangular implants. Experiments by Dunn and Heath[119] on isolated chick heart fibroblasts showed that cytoplasmic microfilaments responded to the radius of curvature of the substrate on which the cells rested. Encapsulation is a normal phenomenon accompanying implantation of solid materials; as porosity and cell ingrowth increase, concentricity of capsular fibers decreases. At certain pore depths and radii of curvature, giant cell formation is stimulated[120] while at other porosities, susceptible materials become calcified.[85]

## A. Cardiovascular Prostheses

Cardiovascular implants are a class in which the distinction between a collagen product and a graft is somewhat obscure. Since the subject is covered in depth elsewhere in this volume, it will be given only brief mention here. In general, any vascular prosthesis, whether reconstituted from purified collagen, a collagenous remnant of a tubular tissue, or an allo-plastic material, must meet common criteria.[121] Purely circumferential or spiral orientation of reinforcements or collagen fibers produces incongruity in effective elastic modulus between graft and host. Radial expansion under pressure is limited by the intrinsic stiffness of the fiber in the circumferential case, and by length contraction and suture strength in the spiral case. Turbulence and thrombosis are common consequences of abrupt change in arterial diameter. The ideal graft would accomplish compliance matching by restoring the elastic component and introducing crimping or relaxation in the collagen component.

Synthetic polymers are notably poor in compliance matching; Dacron has a modulus of $7.3 \times 10^9$ dyn/cm², compared to $3.5 \times 10^6$ dyn/cm² for human ascending aorta.[45,121] Highly cross-linked natural tissues have modulus incompatibilities ranging from 2:1 to 5:1 compared to fresh tissues.[58] The impact of compliance and suturing is even more crucial for diameters under 6 mm.[122]

The deficiencies of simply extracting and tanning small animal vessels, particularly aneurysm consequent to long-term collagenolysis,[122] prompted Ketharanathan and co-workers[123] to investigate methods of forming stronger collagenous tubes, lacking the elastic component, by embedding 0.8 to 1.0 mm diameter silicone mandrels intramuscularly in sheep, and later excising the fibrous capsule. After glutaraldehyde treatment (2.5%, 72 hr) and saline washing, but no other extraction, the grafts were stored in 50% ethanol. Clinical results on larger versions have shown constant 50% patency between 1 and 2 years.[124]

Another approach to stengthening vascular protheses is the use of an alloplastic mesh reinforcement, either as a support for a dried collagen dispersion[125] or as a substrate for in vivo collagen deposition.[126] These methods suffer from the foreign body reaction incited by synthetics such as Dacron, which leads to enhanced collagen turnover and leakage.[127]

Reduction of thrombogenicity is a concern; although high patency rates are claimed for "Ovine" mandrel-grown grafts,[124] there is a high short-term loss rate probably due to thrombosis before growth of a neointima. Methods for reducing platelet adhesion and subseqent thrombosis center on increasing the negative surface charge by controlled cross-linking,[60] by succinylation,[126] or by heparin binding, either ionically, using protamine as an intermediary,[128] or covalently, using carbodiimide cross-linking.[129] The latter resulted in 90% permanently bound heparin at acid pH, which reduced swelling in acid and increased clotting time 14-fold compared to untreated collagen.

It is important that the capacity of collagen for inducing calcification be taken into account in cardiovascular applications, since this mileau is subject to spontaneous calcification at damaged or acellular sites. The high incidence of calcification in heart valve xenografts is a case in point.[130] Calcification of glutaraldehyde-treated heart valves occurs in nonvascular subcutaneous sites, originating at the junction of fibrosa and spongiosa, and intracellularly in fixed donor cell remnants.[131] Calcification is a noncell-mediated process, as indicated by implantation in diffusion chambers; calcium uptake also varied with age and species, being greatest in young rats.[131] Glycerol treatment[132] reduced incidence of calcification and detergent treatment (C12 alkyl sulfate)[133] prevented it, perhaps by interfering with infiltration of osteonectin,[131] but more likely by removal of lipid and cell debris.

## B. Hemostatic Agents

In contrast to the requirements for vascular protheses, hemostatic agents ideally are highly thrombogenic. Unaltered collagen appears to possess such a capacity, interacting directly with platelets[134] which adhere and develop morphological changes indicative of release of phospholipid clotting factor.[135] This knowledge led to the development of "microcrystalline" collagen powder ("Avitene", Alcon Laboratories, Fort Worth, Texas), produced by comminuition of bovine corium with relatively little extraction thereafter. It has a history of adequate performance[136] and may serve as a mucosal dressing as well as a hemostat.[137] However, its powder form and electrostatic properties make positioning the material difficult;[138] scattered particles cause local inflammation and foreign body reaction (as occurs at the treated site during collagenolysis) with the possibility of inducing adhesions.

Silverstein and Chvapil[138] reviewed various hemostatic agents and presented evidence suggesting that an ethylene-oxide-sterilized pad or fleece of mechanically felted 3 to 10 cm long intact fibers extracted from bovine skin possessed superior handling characteristics, although without significantly improved hemostasis.[139] Another alternative is a sponge ("Collastat", Helitrex, Inc., Princeton, N.J.; Figure 5) prepared from bovine tendon dispersion, which showed improved hemostasis and reduced foreign body giant cell population.[140] Sawyer's group[141] reported that "Superstat", a freeze-dried preparation of collagen polypeptides combined with high calcium concentration, reduced bleeding time four-fold compared to Avitene and was dissolved within a fraction of the turnover time for collagen, but lacked sufficient strength for suturing to a wound. Although the extent of intentional denaturation was unclear, it would appear that this material takes advantage of the chemotactic properties of collagen breakdown products and/or ionic gradients to enhance cell ingrowth.

## C. Musculoskeletal Implants

Repair of cystic bone defects by implantation of collagenous xenograft is perhaps the oldest clinical application of collagen. Senn, in 1889,[142] conducted a series of implantations of demineralized bovine bone chips disinfected with iodoform and mercuric alcohol, combined with a study of the same material in trephine and tibial defects in dogs. Senn's theory was that demineralized bone matrix provided a firmer substrate and better separation from surrounding soft tissue than a blood clot alone. This view is still valid, but has been overshadowed by the discovery that a noncollagenous component was capable of inducing nonbone cells to initiate endochondral osteogenesis.[32,143] Since the topic of osteoinduction is dealt with elsewhere in this volume, this section will concentrate on osteoconductive substrates, which require a pool of existing osteoprogenitor cells in immediate proximity.

Collagen implants are manufactured by demineralization of whole or pulverized bone, generally accompanied by lipid extraction, or by reconstitution of dispersed or solubilized collagen into a suitable form, or by recombination of a mineral phase with either of the former organic matrices. Preparation of osteoconductive material is an extension of the procedure for making inductive bone matrix. The distinction should be made between such preparations and other bone-derived implants such as "Kielbone", which is deproteinized rather than demineralized.

A typical process involves: (1) grinding of bovine cortical bone to 74 to 420 μm diameter with liquid nitrogen cooling to prevent thermal denaturation, (2) water washing, (3) lipid extraction using ethanol/ether, chloroform/methanol, acetone, or a detergent, (4) demineralization in 0.5 $N$ HCl for 12 to 48 hr at 4°C (large pieces of bone require longer demineralization), (5) washing to remove acid and dissolved salts, (6) enzyme digestion or dissociative extraction with 4 $M$ guanadine, 8 $M$ urea, or 1% sodium dodecyl sulfate, (7) 0.15 $M$ 2-mercaptoethanol to remove disulfide-bonded noncollagen proteins, (8) final washing or dialysis, and (9) dehydration using hydrophilic solvents or lyophilization.[144] Collagen Corporation, among others, has an active program to develop products based on extraction protocols of this general format. If the starting material were appropriate, the product after step 5 should be osteoinductive; in fact, subcutaneous implants of such materials show a proportion of noninductive particles, particularly at the periphery, which may lack a sufficient concentration of the osteogenic factor.[112] Implanted in a subperiosteal site, such materials may serve as conductive substrates (Figure 6).

Clinically successful implants of demineralized bone are seldom reexamined histologically, and, if excised or biopsied, not necessarily demonstrate endochondral osteogenesis char-

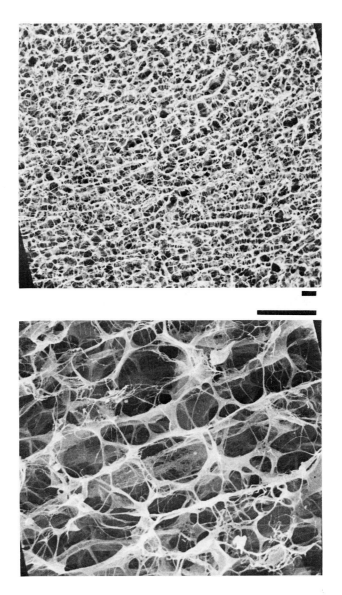

FIGURE 5. Scanning electron micrograph of "Collastat" absorbable collagen hemostatic sponge. Scale bar = 100 μm. Illustration courtesy of J. Pachence, Helitrex, Inc., Princeton, N.J.

acteristic of induction.[145] All particles of demineralized bone are not resorbed during the initial inflammatory phase (4 to 7 days in allogeneic rat implants), and are not ordinarily remineralized.[146] The cause of the resistance of demineralized bone to resorption compared to mineralized bone[147] may relate to lack of binding of macrophages to the demineralized product.[148] Other factors not presently understood may reverse this immunity and stimulate massive foreign body giant cell attack. Yakagi et al.[149] reported this result in implants which were borohydride reduced, not lipid extracted, and enclosed in nylon mesh pouches.

Processes which do not involve solubilization of the matrix have a disadvantage in the inability to remove foreign particles, either of organic origin or from contamination by manufacturing equipment.[112] Assurance of removal of cell debris is uncertain, since cell remnants are identifiable only by histology.[150] Solvent lipid extraction may fix cytoplasmic

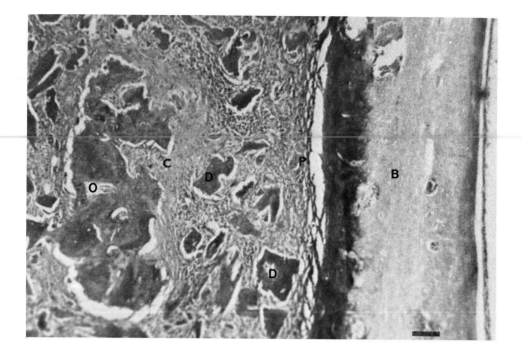

FIGURE 6.    Trichrome-stained 6-μm section of bovine demineralized bone powder after 28-day implantation supraperiosteally under a temporalis muscle flap lateral to the crista frontalis of the parietal in rat. O = Induced bone ossicle; C = capsule; D = demineralized bone particles embedded in periosteum; B = existing parietal bone sectioned perpendicular to surface. Scale bar = 100 μm. (From Sabelman, E. E., Armstrong, R., and Seyedin, S., *Proc. 9th Meet. Soc. Biomater.*, 6, 22, 1983. With permission.)

and nuclear macromolecules; diffusion of substances such as DNA is poor in any case, and such contaminants must be assumed to be present up to a few tenths of a percent, with unknown effects on efficacy. These caveats also apply to intact bone bank and nonbone tissue grafts as well as to purified bone matrix.

Collagen from nonbone sources has received attention as a potential scaffold for osseous repair. DeVore[151] reported on the use of reconstituted bovine skin collagen cross-linked to varying degrees with acrolein, glyoxal, glutaraldehyde, or formaldehyde to repair rabbit mandibular osteotomies. The material was impenetrable to cells and required degradation before replacement by new bone; the greater the degree of cross-linking, the slower was the rate of bone ingrowth (no ingrowth in 12 months in some glutaraldehyde-treated implants, whereas repair had begun by 2 months in unimplanted controls). Formaldehyde-cross-linked material was eroded in 5 weeks, with replacement complete in 3 months; other materials were encapsulated, with foci of new bone isolated from the walls of the defect.

There have been several reports of the use of collagen recombined with bone mineral as potential implantable materials. Gross and colleagues[152] implanted cylindrical plugs of bovine skin collagen gel reacted with $CaCl_2/K_2HPO_4$ subperiosteally adjacent to the rhesus mandible. The implants were not colonized by cells; in some the collagen calcified. In 5 to 7 months periosteal reaction had partially surrounded the implants with new bone; the incidence of exostosis was less with collagen alone (50 vs. 80%) and nil with "Proplast" implants. Miyata et al.[153] described a composite consisting of baked anorganic whole bone vacuum impregnated with atelopeptide collagen; no implantation results were reported. In view of the difficulty of controlling porosity, the inability to restore the original molecular bonding between collagen and mineral, and the possible stimulation of collagenolysis,[148] composites of this type should be extensively tested before conclusions are drawn regarding their efficacy.

Joos and co-workers[154] found that porcine collagen fleece of the same type used by Silverstein and Chvapil[138] would accelerate fibroblast and osteoblast ingrowth by 2 to 3 weeks in rabbit mandibular defects. Earlier, Cucin et al.[155] had found that radiation sterilized (1.5 Mrad) proctase solubilized bovine skin collagen supported callus formation in rabbit rib defects by 2 weeks; however, 30 to 40% of the implant was resorbed. Denatured collagen ("Gelfoam") is resorbed too rapidly to have any effect on bone repair.[156]

**Articular surface repair** — Regeneration of articular cartilage surfaces would be a particularly desirable application of collagenous implants, in view of the widespread incidence of arthritic damage. Spontaneous repair of joint cartilage occurs only if the underlying subchondral cortex is breached, and then produces dysfunctional fibrous tissue with impaired dermatan sulfate synthesis.[157] Narang and Wells[158] replaced mandibular condyles of infant rats and fibular heads of adult rats with allogeneic demineralized bone; mandible length was partially restored by 9 weeks, and disorganized epiphyseal and articular cartilage was regenerated, although surface contour was imperfect. Speer and colleagues[100] implanted glutaraldehyde-cross-linked ethylene oxide-sterilized bovine skin collagen sponge (400-μm average pore diameter) in 6-mm diameter defects in the distal femoral articular surface of rabbits. Sponge-restored surfaces had less contour defect than polyvinyl alcohol sponge or unimplanted controls, developed subchondral cortical bone rather than an encapsulating shell, and had smooth functional junctions with surrounding tissue.

**Oral applications** — Shoshan and co-workers[159] have experimented with homogenized reconstituted collagen mixed with cell culture medium (M-199) for burn treatment and for endodontal repair. The latter consisted of allogeneic collagen in pulpotomies in dogs, with the result that at 1 month inflammation had subsided, debris and implanted collagen had been resorbed, and pulp regenerated. Nevins[107] developed a compound of reconstituted calf skin collagen (up to 60 mg/mℓ) calcium, phosphate, and 0.05% Lugol's solution (5% KI + 10% I) for endodontic canal filling; at 12 weeks cementum-like tissue filled the apical 3 to 4 mm of the canal, while CaOH-filled canals were simply sealed with bony tissue.

**Tendon and ligament** — Collagen is one of the materials under investigation for tendon and ligament repair, both in the form of intact xenografts[40] and as reconstituted colinear fiber tapes.[39] There is concern that collagenolysis during healing may proceed at a greater rate than collagen synthesis, resulting in premature failure. Hence, collagen is being tested as a composite with carbon fiber, replacing polylactic acid.[160] The function of the glutaraldehyde-cross-linked (3%) collagen coating is to delimit cell ingrowth, reduce stress concentration in carbon fibers, and prevent migration of carbon fragments into the joint cavity and synovium (Figure 7). The gradual fragmentation of carbon fiber probably reflects bending fatigue effects; giant cells seen at the ends of fragments most likely form in response to geometric stimuli, rather than cause the breakage. Highly cross-linked colinear collagen might be expected to suffer the same fate, unless replaced by newly synthesized fibers possessing a crimped configuration and elastin component.[44]

### D. Burn and Wound Dressings

Chvapil[161] recently reviewed the arguments favoring collagen as a wound dressing. The main functions of a dressing are to control evaporation, to protect against continued trauma and contamination, and if possible, to reduce existing bacterial count, to permit application of medication, and to encourage regeneration of normal dermal structure, while minimizing scar formation and contracture. Collagen in a single form cannot perform all necessary functions; e.g., a membrane may be an effective moisture barrier but lack adequate adhesion. Accordingly, porous collagen, providing a substrate for tissue repair, is generally used with a laminated synthetic polymer surface.

Chvapil[161] concludes that a collagen sponge having a pore size of 80 μm or more is ideal for ingrowth of fibroblasts and capillaries, combined with a self-adhesive film and polyester

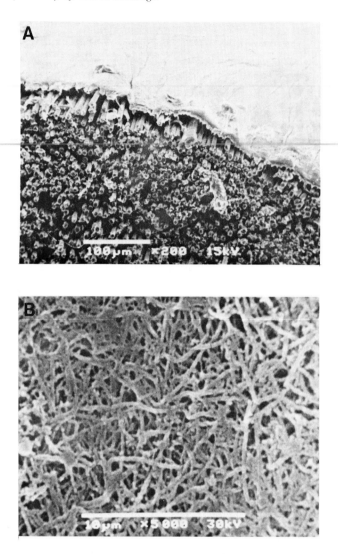

FIGURE 7.   Scanning electron micrograph of carbon fiber ligament prosthesis coated with 15 to 20-μm layer of "Vitrogen 100" atelopeptide
bovine skin collagen (Collagen Corp., Palo Alto, Calif.), air-dried and 3%
glutaraldehyde cross-linked. (A) Cut end of fiber bundle; (B) perpendicular
view of collagen coating. Illustration courtesy of J. L. Ricci, University
of Medicine and Dentistry of New Jersey, Section of Orthopedic Surgery,
Princeton, N.J.

gauze outer lamina. Completely impermeable outer membranes result in local edema, increased sepsis, and decreased reepithelialization by 38% compared to collagen alone. Cross-
linking is necessary, preferably by diisocyanate rather than glutaraldehyde,[80] to reduce
gelation and bacterial growth. Although it was not included as part of the sponge composite,
Chvapil[161] noted that chondroitin sulfate, a glycosaminoglycan, infused into a healing wound
increased its transverse strength by a factor of two.

Yannas and Burke[162,163] have taken a slightly different approach, employing a microper-
forated silicone moisture barrier with coprecipitated, nonbanded bovine skin collagen dispersion and chondroitin-6-sulfate from shark cartilage. The silicone layer is exteriorized by
neoepithelial undergrowth and is replaced by an autograft at 14 to 46 days in the original

procedure. Degradation rate of the collagenous layer is controlled by dry vacuum thermal (105°C) treatment before coating with silicone, and by glutaraldehyde cross-linking in acetic acid after curing the silicone. Cell ingrowth is determined by pore fraction and diameter, which are a function of glycosaminoglycan content, and are ideally 95% and 50 μm, respectively.

To avoid the need for large-area autografting, autologous basal cells isolated within 4 hr of implantation are inoculated by centrifugation into the collagen/GAG layer to create a "Stage 2" skin prosthesis.[164] Scarring, contraction, and time for cell immigration from wound edges were reduced compared to noncell-seeded implants. Efficacy of either version in infected wounds was not reported. Bell et al.[165] also incorporated autologous cells into a rat collagen implant in guinea pigs, but permitted 5 to 7 days proliferation from explants in culture to populate the collagen gel. The proliferating cells were fibroblasts, but their retention in the graft could not be verified after 5 weeks, nor could the source of epidermal cells be traced.

Yannas and colleagues[164] report that significant adhesion of the skin prosthesis to the wound bed occurs in a few hours. This finding may relate to the conclusion of Morykwas and co-workers[166] that zeta potential is a significant indicator of short-term wound adhesion. Positive zeta potential corresponded to increased perpendicular pull-off strength of a 6-mm diameter graft in rats at 5 hr, during which period adhesion was largely fibrin dependent. Uncross-linked porcine skin had $Z = +12.5$ V and supported 85.5 g/cm$^2$ stress, while 10% glutaraldehyde cross-linking resulted in $Z = -26.4$ V and 42 g/cm$^2$ strength. Quenching of residual aldehyde with amino acids reversed both Z and strength trends. Results of long-term (2 month) adhesion tests by Shimizu et al.[103] corroborate the inverse correlation of glutaraldehyde (although not ultraviolet) cross-linking and strength.

Another form of implant for cutaneous reconstruction is the mildly glutaraldehyde-treated, trypsin-digested intact dermal collagen framework under study by Oliver and colleagues.[62] The process (Table 2) has been used for allografts in the rat, with successful revascularization and cell repopulation at 8 1/2 weeks, and for porcine xenografts in rats, with 85% of the original collagen retained, compared to 35% for noncross-linked grafts at 56 weeks. The same process was used experimentally by Griffiths and Shakespeare[113] for human allografts, with biopsies at 1 to 36 months indicating a lack of resorption or encapsulation, and colonization by fibroblasts and capillaries by 3 months. The only dystrophic calcification was at sites of resorbable sutures. The extremely long enzymatic digestion is probably responsible for the success of this material, since it assures extraction of noncollagenous proteins. Oliver's work is reported in detail elsewhere in this volume.

## E. Injectable Collagen Suspensions

One collagen-based material which is becoming accepted into routine clinical practice is Zyderm Collagen Implant (Collagen Corporation, Palo Alto, Calif.), prepared from pepsin-digested reconstituted bovine skin collagen by a modification of the method of Knapp et al.[77] The material is supplied in syringes at concentrations of 35 and 65 mg/mℓ in phosphate-buffered saline, ordinarily compounded with 0.3% lidocaine, for use intradermally for correction of facial contour irregularities.[167] Efficacy is technique-dependent, in that material injected subdermally in humans is rapidly dispersed and resorbed, and lesions must be overcorrected to compensate for syneresis of fluid during the first 24 hr and later condensation under the pressure of surrounding tissues, to a final volume 25 to 40% of its initial bulk. Correction is most stable if the source of the lesion was nonrecurring, such as trauma or acne scarring, and persistence is shortest when the skin is subject to continued stress.[168,170]

The prospective patient is required to undergo a 0.1-cc test dose 1 month before full treatment with Zyderm implant. During clinical trials, 3% of patients had untoward responses, 81% of which were confined to the test site; cause of presensitization is usually unknown, but may include immune response to lidocaine as well as to bovine collagen. Of adverse reactions in treated patients ascribable to the implant, rather than to injection artifacts, about half were due to unrecognized positive test responses and half to acquired sensitivity during the course of treatment.[169] Patients since the end of clinical trials have experienced the same proportion of adverse reactions, consisting of granulomas, with foci of eosinophilia in the collagen of the implant and infiltration of histiocytes and giant cells at positive test sites[171] and systemic arthralgia or localized erythematous nodules at treatment sites.[172] There is no reason not to expect a similar incidence of immunogenic reactions to other forms of implantable collagen; Zyderm implant is the first to have a patient population of sufficient size to accumulate statistically valid data.

Laboratory testing of Collagen Corporation's product is by suprascapular subdermal injection of 1-cc doses in rats.[77] The response usually consists of peripheral fibroblast immigration by day 5, fibroblast penetration to half the implant diameter, and increased vascularity by day 15 (Figures 7 and 8), and complete colonization with peripheral adipocytes by day 30.[173] Cell penetration occurs largely along planes of low fiber density; if the material is entirely homogeneous and of high collagen concentration, immigration is impeded. Histology of human biopsies varies substantially from the laboratory case, due to fragmentation of the implant by flow along preexisting tissue planes, as well as mechanically induced compaction (Figure 9). The result after several months does not differ greatly from that described by Griffiths and Shakespeare,[113] despite the absence of cross-linking.

Other applications have been suggested for injectable collagen suspensions, such as intravascular occlusion for treatment of inoperable tumors[174] and augmentation of the esophageal sphincter to prevent reflux. A similar, albeit noninjectable, material ("Collafilm" gel) was used by Marzin and Rouveix[175] for stimulation of granulation in chronic leg ulcers, with reduction of original wound area to 12%, compared to 37% for conventional treatment at 12 weeks.

## F. Immobilized Enzymes

Collagen has potential for both medical and industrial application as a biocompatible low-cost substrate for immobilized enzymes. Klibanov[176] reviewed industrial methodology and economics and noted the problems of bound enzymes with regard to reagent partitioning and local nonoptimal pH, diffusional limitations, and steric hindrance, which lower activity relative to free enzyme although long-term stability is improved. Experiments with collagen-bound enzymes have shown that diffusional reduction[177] and steric hindrance[178] also apply in this specific case.

Bernath and co-workers[179] discussed medical applications and described extracorporeal apparatus necessary to perfuse a patient's blood through an immobilized enzyme reactor. The spiral membrane reactor used by Bernath's group caused a recoverable 24% drop in platelet count during 2 hr perfusion, and a 48% loss in leukocyte count during the 1st hr, followed by rebound to 150% of normal; the latter was due to adsorption on noncollagen components of the flow loop.

Starting material is either dried collagen film, similar to the dialysis membrane of Stenzel et al.[82] or acid dispersions (Table 5). Methods for binding the enzyme vary from glutaraldehyde cross-linking[179,180,182] to coelectrodeposition.[76] Wang and Vieth[181] also employed electrodeposition, and pointed out that enzyme-collagen binding occurs by a combination of salt linkage, hydrogen bonding, and Van der Waals interaction, without the need for aldehyde cross-linking. Lee and others in Coulet's group[178] compared coupling of enzymes to collagen film by the use of Woodward's "K" reagent, carbodiimide, and the acyl-azide technique, and concluded that carbodiimide retained equivalent activity but was superior in permitting activation of binding sites to precede addition of the enzyme. This reduced denaturation of the enzyme caused by long exposure to reagent "K" or glutaraldehyde;

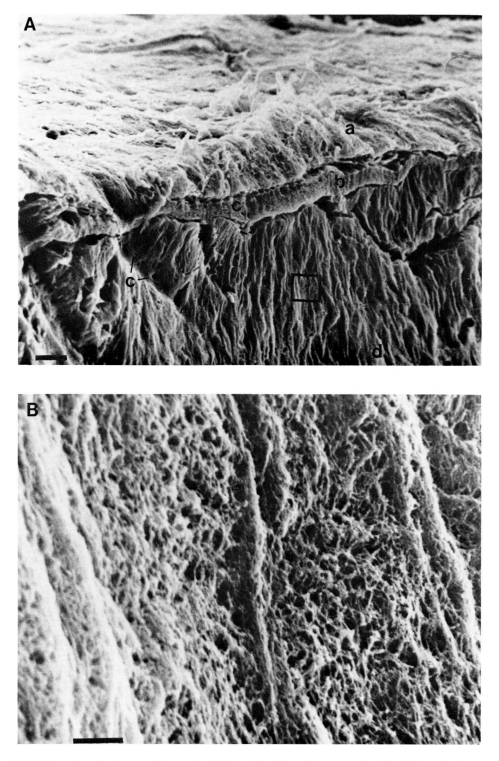

FIGURE 8. Scanning electron micrograph of "Zyderm" collagen fiber suspension after 14-day subcutaneous implantation in rat. (A) Overview of cleaved surface. a = host connective tissue bed; b = capsule; c = cell lacunae in capsule and implant; d = uncolonized interior of implant. Scale bar = 100 μm. Figure 8B is a magnified view of the boxed area in (A), showing acellular fiber structure. Scale bar = 10 μm. (From Armstrong, R., Cooperman, L. S., Parkinson, T. A., and Piez, K. A., *Contemporary Biomaterials,* Boretos, J. W. and Eden, M., Eds., Noyes, Park Ridge, N.J., 1984. With permission.)

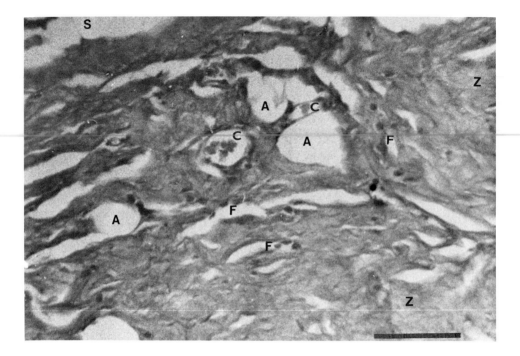

FIGURE 9.   Trichrome-stained 6-μm section of ''Zyderm'' collagen fiber suspension after 14 day-subcutaneous implantation in rat. (A) adipocytes; (C) capillaries; (F) fibroblast clusters; (S) implant surface; (Z) acellular implant material. Scale bar = ca. 100 μm. (From Armstrong, R., Cooperman, L. S., Parkinson, T. A., and Piez, K. A., *Contemporary Biomaterials*, Boretos, J. W. and Eden, M., Eds., Noyes, Park Ridge, N.J., 1984, chap. 25. With permission.)

denaturation could also be reduced by including ATP, dithiothreitol, creatine, or glucose during coupling.

Watanabe et al.[183] described a method for reducing the volume of collagen required and improving handling by reinforcement with polyethylene or polypropylene mesh or yarn. Surface free radicals were generated on the synthetic polymer by plasma discharge, then the activated material was immersed in 1% collagen solution at pH 3, which was neutralized with ammonia, dried, cross-linked in 2.5% glutaraldehyde for 10 min, washed in methanol, and finally, coupled to urokinase or trypsin by the acyl-azide method. Bound urokinase was found to be less sensitive to bulk pH changes than free enzyme.

Karube et al.[182] made the interesting discovery that lipase activity could be regulated by an electric field or current, and that inclusion of a liquid crystal (4-methoxybenzilidene-4'-N-butylaniline) in the collagen-enzyme composite enhanced this effect. Specific activity increased 3.4 times when 4 V were applied between the collagen membrane bonded to a platinum sheet cathode and a platinum anode 0.3 cm distant; without the liquid crystal the increase was less than double. Sensitivity to pH was also shifted, suggesting that the liquid crystal dipole affects local electric field and reagent mobility; measurement of the zeta potential could perhaps clarify this question.

## G. Drug Delivery Devices

Closely related to immobilization of enzymes is the use of collagen for delivery of drugs, an application making use of the gradual solubilization of the matrix and diffusional loss of admixed material both of which are disadvantageous in the former case. The earliest inclusion of collagen in drug delivery systems was as an emulsion stabilizer for oral, nasal, and vaginal administration.[81] Other more recent proposals for collagen sponge vaginal inserts for delivery

**Table 5**
**IMMOBILIZED ENZYMES**

| Enzyme | Method | Result | Ref. |
|---|---|---|---|
| Glucoamylase | Mix enzyme + dispersion, pH 7, dry in sheet, 10% glutaraldehyde, for 3 min | 2/3 activity after 400 min.; loses 21% in storage 200 days | 180 |
| L-Asparaginase | Mix enzyme + lactic acid dispersion, pH 4.2, dry, 1% glutaraldehyde, 30 sec, pH 7 | Stable at 2/3 initial activity after 1 to 7 uses | 179 |
| Aspartate aminotransferase | 1. Collagen film swollen in 0.1 $N$ | Activity: EDC = 2 × "K" | 178 |
| Creatinine kinase | NaOH, 2 hr @ 37°C | EDC 30% better than azide | |
| Lactate dehydrogenase | 2. a. Mix 2 mg/m$\ell$ enzyme + | "K" best | |
| Malate dehydrogenase | 0.08 $M$ Woodward's "K" | No difference | |
| Hexokinase | in borate buffer, pH 8.7 | EDC = 1.5 × azide | |
| Urease | b. Mix 2 mg/m$\ell$ enzyme + 0.1 M EDC in phosphate buffer, pH 8 | Azide best | |
| | c. Activate in 0.1 $M$ EDC, 6 hr @ 4°C, pH 4.6, then couple enzyme, 1—4 mg/m$\ell$ in borate buffer, pH 8.7 | | |
| | 3. React 24 hr @ 4°C | | |
| | 4. Wash in phosphate buffer, pH 7.4 | | |
| Malate dehydrogenase | Acyl-azide | 40% of free enzyme activity, nonlinear with concentration | 177 |
| Invertase | Impregnation or electrodeposition | | 181 |
| Lysozyme | | Stable activity = 35% | |
| Urease | | Stable activity = 25% | |
| Glucose oxidase | | Stable activity = 15% | |
| Glucose isomerase | | | |
| Penicillin amidase | | | |

*Note:* EDC = 1-ethyl-3(3-dimethylaminopropyl) carbodiimide HCl.

of retinoic acid have suffered from lack of control of delivery rate and high retention of the drug within the collagen sponge.[184] This deficiency is probably due to the lack of mixing of collagen and drug on the molecular level, the material simply coating the trabeculae of the sponge.

Collagen has been used in ocular therapy in several instances. Rubin et al.[185] made dried films and glutaraldehyde cross-linked (0.5%, pH 8.5, 18.5 hr) gels from atelopeptide collagen; 4% pilocarpine was added to films before drying, and to gels by rehydration in 50% pilocarpine before cross-linking. Release of 100% of the highly soluble pilocarpine was delayed by 8 to 10 minutes from gels and 25 min from membranes, and was entirely determined by dissolution of the implant in rabbit eye. The authors suggested reducing the positive charge of collagen by succinylation to enhance electrostatic binding and so delay release. Conjunctival inserts of succinylated collagen made by this method were combined with gentamicin by Slatter et al.[186] for treatment of bovine keratoconjunctivitis. Problems were encountered in optimizing dimensions of the implant to achieve adequate retention time. In vivo measurement showed that the level of drug in surrounding fluid declined 95% in 2 hr but remained at therapeutic levels for 16 hr.

## H. Neurosensory Applications

Repair of sensory and neural dysfunction is a category in which collagen portends to make a significant contribution. One simple application is a disposable contact lens based on a patent of Miyata which is said to be nearing clinical test.[108]

Otologic repair has seen considerable experimentation in the use of collagen xenograft tympanic membrane[187] and ossicular[188] replacement. In the latter example, collagen sponge replaced Gelfoam in Schulknecht or Armstrong-type prosthetic stapes. More recently, Liston[189] used compressed Avitene to seal oval windows following stapedectomy in cats, and reported that improved hemostasis should reduce the incidence of granulomas complicating Gelfoam-and-wire implants. Since only small amounts of implant material are needed for such applications, autografts will probably retain a major share of routine auricular reconstruction, with some use being made of allografts.[190]

Many attempts have been made to improve posttraumatic peripheral nerve regeneration, given that epineural suturing of severed nerve ends results in a low incidence of functional restoration. Early attempts to approximate the ends using resorbable collagen cuffs or wrappings were not notably successful, but eliminated the need for reentrance to remove silicone cuffs.[191] In a recent study, Rosen et al.[192] suggested that fascicular (perineural) suturing improved alignment of severed or crushed fibers in rat saphenous nerves, but that atelopeptide collagen membrane perineural sheaths produced superior integrated monophasic compound action potentials at 12 to 24 weeks (82% of normal vs. 43% sutured vs. 50% fibrin clot alone). The wrapping method employed was not capable of supporting tensile loads, and would have to be augmented if the nerve were under tension.

## I. Collagen as a Cell Culture Substrate

Since 1956,[193] collagen has been used as an adjunct to culture of isolated cells, both to avoid the artifacts associated with monolayer culture on glass and plastic, and to investigate the relationship between cells and extracellular matrix.[13] While the literature is too large to be thoroughly reviewed here, this section will examine variations in methodology and their appropriateness for specific cell types and functions (Table 6).

The starting material for cell substrates has nearly always been rat tail tendon in acid solution, and the endproduct an air-dried coating on petri dishes or glass cover slips, usually preneutralized and kept cold, or a gel made by pouring neutralized solution into dishes (Figures 2 and 3). If not filter-sterilized, the material is often UV-sterilized under a laminar flow hood (Table 4).

Elsdale and Bard[204] developed a method that has since been often cited: cold 0.1% HCl solution of rat tail tendon (Table 3) at pH 4 is rapidly mixed with 10% fetal calf serum, 10X concentrated culture medium, such as Eagle's MEM, and 0.14 $M$ NaOH to achieve pH 7.6. The mixture is kept on ice until poured and then gelled at room temperature, with a turbidity plateau after 15 to 30 min. When poured over an existing fibroblast layer, the cells will detach from plastic and migrate into the gel, assuming a spindle-shape characteristic of the in vivo phenotype. Nonattachment-dependent neoplastic cells behave in collagen gels much the same as in agar suspension, forming isolated clones. Many cell strains were slower to start proliferating than on plastic, a finding verified by others.[73,208]

Elsdale and Bard[204] found that fibroblasts were not released from the gel by trypsin alone, contrasting with the behavior of transformed cell lines[200] and suggesting that fibroblasts become enmeshed in newly synthesized collagen. A further finding of Elsdale and Bard was that collagen fibril bundles could be aligned by tilting or draining the dish before the turbidity maximum. Dunn and Ebendal,[89] using this technique, found that fibroblasts were highly colinear with the matrix, but that contact guidance was lost if the gels were dried and collapsed.

Several experiments have compared cell behavior on or within gels with that on dried collagen coatings. Kleinmann et al.[205] found that the optimum substrate for Chinese hamster ovary cells was a coating dried from 10 $\mu$g/m$\ell$ collagen gel, since only 5% serum provided

**Table 6**
# COLLAGEN SUBSTRATES FOR CELL CULTURE (TYPE I UNLESS NOTED)

| Cell type | Variables | Result | Ref. |
|---|---|---|---|
| Human foreskin fibroblast | Attached gel | Contraction faster if cut | 194 |
| | Floating gel | Contracted to 20% diameter | |
| Mouse mammary end bud | In gel | Basal lamina regenerates | 195 |
| Hamster renal tumor | Monolayer | 90% cell loss in 10 days under | 196 |
| | In floating gel | 50% contraction, no cell loss | |
| Chick heart fibroblast | Aligned wet gel | Increased cell migration | 89 |
| | Aligned dried gel | | |
| | Dried, rehydrated | Depth not restored | |
| | Thick gel on cover glass | | |
| Xenopus and chick myoblast, mesenchymal, epithelial cells | Intact Xenopus tail, trypsin-digested | Epithelial cells do not penetrate | 12 |
| Guinea pig dermal fibroblast, epidermal cells | Dried Type I, rat | 90% fibroblasts attach, require | 11 |
| | Dried Type II, sarcoma | fibronectin; 25% epidermal | |
| | Dried Type III, fetal calf | cells attach only to Type IV, | |
| | Dried Type IV, sarcoma | without fibronectin | |
| Rat epithelial tumor | Floating rafts: | | 13 |
| | Type I, rat tail | | |
| | Type III, rat skin | | |
| | Type IV, rat skin | | |
| Adult bovine aortic endothelium | In gel | Neocapillaries under gel | 197 |
| | Gelatin coating | | |
| Neutrophil granulocytes | Coated glass | No adhesion or penetration | 198 |
| | In gel, 1.0 mg/m$\ell$ | Penetration to 270 $\mu$m | |
| | In gel, 2.5 mg/m$\ell$ | Penetration to 120 $\mu$m | |
| Chick sternal chondrocytes | On plastic dish | Higher population per dish | 199 |
| | In gel | New low MW collagen | |
| HeLa, BHK cell lines | Heat denatured film | Serum improves attachment, | 200 |
| | Urea denatured film | trypsin detaches cells on dena- | |
| | Denatured-dried film | tured films | |
| | Gel | | |
| | Gel + urea | | |
| Chick sympathetic, cerebral neurons | See Table 4 | Attachment and growth vary with cell type and substrate | 86 |
| BHK, human skin fibroblast | Glass | BHK: flat, require fibronectin | 201 |
| | Dried coating | on glass and coating | |
| | Gel | BHK: rounded, no fibronectin | |
| Normal and neoplastic mouse mammary cells | Plastic | 3 X proliferation, stop 3 days | |
| | On gel | 10 X proliferation in 14 days | |
| | In gel | 20 X proliferation | |
| HeLa, BHK, adult human skin fibroblast, RPMI-3460, hamster melanoma | On gel | Human fibroblast lags cell lines; | 73 |
| | In gel | epithelial cells cannot penetrate | |
| | Dried neutral coating | | |
| 3T3 cells | Millipore filter | 50—80% attach on bare filter | 203 |
| | Heat gel on filter | and on gel with fibronectin | |
| | Ammonia gel on filter | | |
| Rat sympathetic neurons | Dried, UV sterilized | Gels, polypeptides best adhe- | 105 |
| | Ammonia gel | sion; killed cells next best | |
| | Photopolymerized | | |
| | Polylysine on gelatin | | |
| | Polyornithine | | |
| | Histone | | |
| | Killed cells (heart) | | |
| Human fetal lung fibroblast, rat fibroblast + SV$_{40}$, BHK21, HeLa | On gel (drain aligned) | Fibroblasts are spindle-shaped | 204 |
| | Under gel | vs. ruffled on plastic; not tryp- | |
| | Mixed in gel | sin detached; SV$_{40}$-transformed | |
| | Floating drops of gel | cells do not attach to gel | |

**Table 6 (continued)**
**COLLAGEN SUBSTRATES FOR CELL CULTURE (TYPE I UNLESS NOTED)**

| Cell type | Variables | Result | Ref. |
|---|---|---|---|
| Chinese hamster ovary | Wet and dried, $+/-$ serum: Ammonia neutralized gel Tris buffer heat gel | | 205 |
| | Phosphate buffer gel HEPES buffer gel | Attachment high, no spreading | |
| Mouse lymphocytes | On gels Dried gels | 55% attach | 206 |
| | Acid-rehydrated dried gel | 35% attach | |
| HeLa, bovine pulmonary artery endothelium | Sulfonated Bio-beads: As is (negative surface) | 70% attach in 10 hr | 207 |
| | Polyethylenimine ($+$) Glutaraldehyde ($+$), NaBH | 100% attach | |
| | Collagen | 85% attach; 90% with fibronectin | |

sufficient fibronectin for 75% attachment. Up to 40% of the dried collagen redissolved in cell-free culture medium after 90 min at 37°C, implying inadequate polymerization. Bullaro[209] described in detail a standard protocol for preparation of room-temperature-dried coated dishes from dialyzed rat tail tendon solution for culture of embryonic muscle monolayers. The study of Iversen et al.[86] showed that drying unneutralized acid solutions produced extremely small disorganized fibrils rather than physiologic fiber bundles (Figure 2); hence, exhaustive dialysis or other neutralization is essential to initiate polymerization. Michalopoulos and Pitot[210] published a study showing that floating hydrated gels supported liver parenchymal cell function much better than dried coatings; however, drying was by exposure of acid solution to 50°C for 40 hr, followed by neutralization, resulting in thermal denaturation that makes the two substrates incomparable.

Several other forms of collagen have been used as cell supports. For example, Daniels[211] grew hematopoietic marrow and McAteer and co-workers[212,213] cultured rat and human lung explants and cells on Gelfoam sponges; these publications claim Gelfoam is composed of collagen, whereas the manufacturer (Upjohn, Kalamazoo, Mich.) describes it as ''gelatin.'' Gelatin may be an adequate support for marrow, since in the procedure of Daniels[211] it is supplemented with 0.3% agar, a common medium for marrow culture.

Two other sponge-like matrixes of limited utility for cell culture are porcine skin, as prepared for burn therapy[214] and collagen-impregnated cellulose sponge.[215] Sponges have the advantage of supporting large tissue-density cell masses, but observation of central regions during culture is not possible and diffusion limitations cause phenotypic differences between interior and exterior cells.[215]

Collagen coatings have been used for surfaces other than Petri dishes, in particular, for gas-permeable Teflon films[216] and Millipore-type cellulose filters.[203] Covalent coating to glass is possible using 3-aminopropyltrimethoxysilane and glutaraldehyde as intermediates,[217] although collagen was not one of the proteins tested in this report. Collagen coating was effective in reducing the negative charge of sulfonated polystyrene microcarrier beads, using carbodiimide or glutaraldehyde as coupling agents.[207]

Achieving specificity of the substrate for a given cell type has been attempted by addition of fibronectin and/or GAGs.[13,201,218] Fibronectin is a component of serum, which is routinely added to culture media, supplying the need of some cells for this attachment intermediary.[203] Even those cells which need fibronectin to bind to denatured[200] or dried[201] collagen have a second serum-independent mechanism for attaching to hydrated native collagen. This mechanism apparently does not distinguish between Types I, II, and III, and does not bind to Types IV and V.[219]

Fibronectin was found to inhibit migration of fibroblasts and stimulate melanoma cells to move into hydrated gels, without affecting adhesion, proliferation, or morphology of either type.[220] In contrast, Murray et al.[221] found adhesion of fibroblasts to Types I and IV collagen to be increased from 20 to 80% by fibronectin, but epidermal cells, of which 8% bound to Type I and 25% to Type IV, were unaffected by fibronectin.

The locomotion and proliferation of cells on or within collagen gels are highly dependent on strain and phenotype. Schor and co-workers reported that proliferation of HeLa cervical carcinoma cells is unaffected by substratum or location, but that fibroblasts proliferate more rapidly on the surface of gels[73] and vascular endothelial cells are polygonal on the surface and elongated, similar in appearance to capillary sprouts, within gels.[197] Locomotion of cells such as neutrophils[198,222] and lymphocytes,[206] which lack the specific collagen binding mechanism, is by insertion and expansion of pseudopodia between collagen fibers, and is therefore inhibited by increasing collagen concentration.

## V. THE ROLE OF COLLAGEN IN MORPHOGENESIS

That the extracellular matrix exerts a strong influence over the course of organismal ontogeny has been known for some time.[32,218,223] In turn, the interactions of cells with the matrix, with local messengers or transforming factors, and with other cells, causes them to alter their secretion of new matrix to meet the demands of the growing organism.[224,225] Clearly, the synthetic activity of the cell must be regulated by contact with surrounding collagenous matrix and must detect changes in mechanical stress and matrix biochemistry, not only for purposes of ontogenesis but also for maintenance and adaptation of the differentiated state.[8,9]

There is ample evidence that cells can physically respond to the orientation of the substrate on which they rest.[119] Simply changing the shape of cells by permitting them to contract without releasing their contacts with the matrix seems to be sufficient to cause modulation of phenotypic synthesis.[226] For example, mouse mammary cells on floating collagen gels begin synthesizing caseins; other proteins remain unaffected by substratum.[227] By including more complex components such as Type IV collagen in the matrix, lactalbumin secretion can be increased tenfold over that on Type I gels.[228] This and similar studies of mammary morphogenesis[195] began with explants, rather than with cells alone. Branching glandular morphogenesis from isolated cells is possible, and occurs only on floating freely contractable gels.[229]

Another example of the influence of attachment and consequent alteration of morphology on differentiation is chondrogenesis by isolated cartilage cells. While such cells will secrete GAGs while plated on plastic dishes, they will elaborate Type II matrix only when cell density[230] or suspension in agarose[231] permit return to a spherical shape. Undifferentiated mesenchymal cells at low densities apparently can bind to Type I collagen and thereafter initiate chondrogenesis.[232] Binding of chondrocytes to collagen gel is highest in the absence of fibronectin, which results in fibroblastic phenotype[199] and in the presence of a serum protein termed "chondronectin".[29] Since chondrocytes normally are not exposed to serum, it remains to be shown that this factor exists in avascular cartilage.

Responses to some diffusible factors that modulate cell function appear to be potentiated by contact with the extracellular matrix. In addition to the mammary cell modulation previously mentioned, capillary endothelial cells initiate proliferation in response to angiogenic factor only when attached to collagen.[233] While fibroblastic cells may respond to diffusible osteogenic factor or bone morphogenic protein, the osteoinductive response appears to be mediated by contact with implanted matrix,[32,143] the first sites of chondrogenesis being in crevices and pockets between demineralized bone particles.[112]

Although some cell-cell interactions are substratum independent, others do not occur unless the extracellular milieu is favorable.[218] Morphogenesis of islet-like organoids is stimulated by culture of pancreatic cells between nondetached collagen gel layers.[234] This event occurs by migration of B cells toward the center of clusters and non-B cells toward the periphery, and hence is not shape dependent. The interaction between Schwann cells and nerve fibers that culminates in formation of the myelin sheath depends on the requirement of the Schwann cell for a collagen substrate.[235]

How is information on the properties of extracellular matrix transmitted to cytoplasmic and nuclear regulating mechanisms? Yamada[9] thoroughly reviewed current knowledge of matrix receptor proteins on the cell membrane. Dodd et al.[194] presented data showing that order (rigidity) of the lipid membrane was increased by matrix interaction. Dunn and Heath[119] showed changes in cytoplasmic microfilaments, and Tomasek et al.[236] demonstrated concentration of actin and alpha-actinin at sites of attachment. Bissell and colleagues[8] drew on this body of knowledge of cytoskeletal responses to the extracellular matrix to formulate an hypothesis of dynamic reciprocity between contacts with matrix and cellular synthetic and genetic apparatus (Figure 1). If the cell in question is one that synthesizes matrix, the result of the interaction is the ordered, directional secretion of new collagen, according to the mechanism postulated by Trelstad et al.[21] which closes the feedback loop in the structural servomechanism of the organism.

## A. Piezoelectricity of Collagen

Piezoelectricity and other electrical properties of collagen have been discussed at length in the literature as possible stress transducers influencing cell function, particularly in bone,[237-239] but also in skin.[240] Although the above hypothesis of direct biochemical linkage between cell and matrix would imply that a piezoelectric transducer is redundant, the fact remains that collagen is piezoelectric and capable of generating electrical potentials under stress by a process of polarization and shifting of hydrogen bonds.[241] Dehydration increases the effect,[242] as does cross-linking,[243] due to closer crystalline packing and reduction of conductivity of the surrounding medium.

Since many inductive interactions are surface-charge dependent, it is possible that piezoelectric charge separation could play a part in the local variation of bone resorption and redeposition.[244] Or the interaction of self-generated voltage gradients with cells attached to a random substrate could polarize the cells and impose a vector on their locomotion and secretion, as apparently occurs in neural crest cells during embryogenesis.[245] These synergistic phenomena suggest the possibility of research toward collagen-based prostheses which will actively encourage functional regeneration at the cellular level.

## ACKNOWLEDGMENTS

The author wishes to thank the following for assistance and advice during preparation of the manuscript: K. Hughes of Battelle-Columbus, K. Piez, B. Weiss, and H. McCleary of the Collagen Corporation, D. Speer and M. Chvapil of the University of Arizona, Lane Library, Stanford University School of Medicine, and those cited elsewhere who provided illustrations.

# REFERENCES

1. **Ramachandran, G. N. and Gould, B. S., Eds.,** *Treatise on Collagen,* Academic Press, New York, 1967—1968.
2. **Eyre, D. R.,** Collagen: molecular diversity in the body's protein scaffold, *Science,* 207, 1315, 1980.
3. **Gay, S. and Miller, E. J.,** The biochemistry and metabolism of collagen, in *Collagen in the Physiology and Pathology of Connective Tissue,* Gustav Fischer Verlag, New York, 1978, chap. 1.
4. **Bornstein, P.,** The biosynthesis of collagen, *Ann. Rev. Biochem.,* 43, 567, 1974.
5. **Burgeson, R. E.,** Genetic heterogeneity of collagens, *J. Invest. Dermatol.,* 79, 25s, 1982.
6. **Sage, H.,** Collagens of basement membranes, *J. Invest. Dermatol.,* 79, 51s, 1982.
7. **Laurie, G. W., Leblond, C. P., and Martin, G. R.,** Localization of type IV collagen, laminin, heparan sulfate proteoglycan and fibronectin to the basal lamina of basement membranes, *J. Cell. Biol.,* 95, 340, 1982.
8. **Bissell, M. J., Hall, H. G., and Parry, G.,** How does the extracellular matrix direct gene expression?, *J. Theor. Biol.,* 99, 31, 1982.
9. **Yamada, K. M.,** Cell surface interactions with extracellular materials, *Ann. Rev. Biochem.,* 52, 761, 1983.
10. **Sugrue, S. P. and Hay, E. D.,** Interaction of embryonic corneal epithelium with exogenous collagen, laminin and fibronectin: role of endogenous protein synthesis, *Devel. Biol.,* 92, 97, 1982.
11. **Murray, J. C., Stingl, G., Kleinman, H. K., Martin, G. R., and Katz, S. I.,** Epidermal cells adhere preferentially to type IV (basement membrane) collagen, *J. Cell. Biol.,* 80, 197, 1979.
12. **Overton, J.,** Response of epithelial and mesenchymal cells to culture on basement lamella observed by scanning microscopy, *Exp. Cell Res.,* 105, 313, 1977.
13. **Reid, L. M. and Rojkind, M.,** New techniques for culturing differentiated cells: reconstituted basement membrane rafts, *Methods Enzymol.,* 63, 263, 1979; U.S. Patent 4,352, 887, 1982.
14. **Campo, R. D. and Tourtellotte, C. D.,** The composition of bovine cartilage and bone, *Biochim. Biophys. Acta,* 141, 614, 1967.
15. **Mow, V. C. and Schoonbeck, J. M.,** Comparison of compressive properties and water, uronic acid and ash contents of bovine nasal, bovine articular and human articular cartilages, *Proc. 28th Annu. Orthop. Res. Soc.,* 209, 1982.
16. **Helspeth, D. L., Lechner, J. H., and Veis, A.,** Role of the amino-terminal extrahelical region of type I collagen in directing the 4D overlap in fibrillogenesis, *Biopolymers,* 18, 3005, 1979.
17. **Piez, K. A.,** Structure and assembly of the native collagen fibril, *Connect. Tissue Res.,* 10, 25, 1982.
18. **Barnes, M. J.,** Biochemistry of collagens from mineralized tissues, in *Hard Tissue Growth, Repair and Remineralization,* Ciba Found. Symp. 11, new ser., 1973, 247.
19. **Tanzer, M. L. and Waite, J. H.,** Collagen cross-linking, *Coll. Rel. Res.,* 177, 1982.
20. **Golub, L., Stern, B., Glimcher, M., and Goldhaber, P.,** The inhibition of the maturation of newly synthesized bone collagen by beta-aminopropionitrile in tissue culture, *Proc. Soc. Exp. Biol. Med.,* 129, 465, 1968.
21. **Trelstad, R. L., Birk, D. E., and Silver, F. H.,** Collagen fibrillogenesis in tissues, in solution and from modeling: a synthesis, *J. Invest. Dermatol.,* 79, 109s, 1982.
22. **Traub, W.,** Molecular assembly in collagen, *FEBS Lett.,* 92, 114, 1978.
23. **Trelstad, R. L. and Silver, F. H.,** Matrix assembly, in *Cell Biology of Extracellular Matrix,* Hay, E., Ed., Plenum Press, New York, 1981, 179.
24. **Lees, S.,** A mixed packing model for bone collagen, *Calcif. Tiss. Int.,* 33, 591, 1981.
25. **Schiffmann, E., Martin, G. R., and Miller, E. J.,** Matrices that calcify, in *Biological Calcification: Cellular and Molecular Aspects,* Schraer, H., Ed., Appleton-Century-Crofts, New York, 1970, 27.
26. **Gerlach, U.,** Metabolism and structure of connective tissue during extraosseous calcification, *Clin. Orthop. Rel. Res.,* 69, 118, 1970.
27. **Baylink, D., Wergedal, J., and Thompson, E.,** Loss of protein-polysaccharides at sites where bone mineralization is initiated, *J. Histochem. Cytochem.,* 20, 279, 1972.
28. **Hauschka, P. V., Lian, J. B., and Gallop, P. M.,** Isolation of a bone protein containing the calcium-binding amino acid gamma-carboxyglutamate, *Proc. 22nd Meet. Orthop. Res. Soc.,* 18, 1976.
29. **Hewitt, A. T., Kleinman, H. K., Pennypacker, J. P., and Martin, G. R.,** Identification of an adhesion factor for chondrocytes, *Proc. Natl. Acad. Sci. U.S.A.,* 77, 385, 1980.
30. **Anderson, H. C.,** Vesicles associated with calcification in the matrix of epiphyseal cartilage, *J. Cell. Biol.,* 41, 59, 1969.
31. **Imai, Y. and Masuhara, E.,** Long-term *in vivo* studies of poly(2-hydroxyethyl methacrylate), *J. Biomed. Mater. Res.,* 16, 609, 1982.
32. **Reddi, A. H.,** Collagen and cell differentiation, in *Biochemistry of Collagen,* Ramachandran, G. N. and Reddi, A. H., Eds., Plenum Press, New York, 1976, chap. 9.
33. **Woodward, S. C.,** Mineralization of connective tissue surrounding implanted devices, *Trans. ASAIO,* 26, 697, 1980.

34. **Lees, S. and Davidson, C. L.,** The role of collagen in the elastic properties of calcified tissues, *J. Biochem.,* 10, 473, 1977.

35. **Crisp, J. D. C.,** Properties of tendon and skin, in *Biomechanics, Its Foundations and Objectives,* Fung, Y. C., Perrone, N., and Anliker, M., Eds., Prentice-Hall, Englewood Cliffs, N.J., 1972, chap. 6.

36. **Grood, E. S., Noyes, F. R., and Miller, E. H.,** Comparative mechanical properties of the medial collateral and capsular structures of the knee, *Proc. 23rd Annu. Orthop. Res. Soc.,* 102, 1977.

37. **Morein, G., Goldgefter, L., Kobyliansky, E., Goldschmidt-Nathan, M., and Nathan, H.,** Changes in the mechanical properties of rat tail tendon during postnatal ontogenesis, *Anat. Embryol.,* 154, 121, 1978.

38. **Haut, R. C.,** Age-dependent influence of strain rate on the tensile failure of rat-tail tendon, *Proc. 29th Meet. Orthop. Res. Soc.,* 85, 1983.

39. **Hughes, K. E., Fink, D. J., Hutson, T. B., and Veis, A.,** Oriented fibrillar collagen and its application to biomedical devices, American Leather Chemists Association, 79th Annual Meeting, June 22, 1983.

40. **Berg, W. S., Stahurski, T. M., Moran, J. M., and Greenwald, A. S.,** Mechanical properties of bovine xenografts, *Proc. 29th Meeting, Orthop. Res. Soc.,* 87, 1983.

41. **Ippolito, E., Natali, P. G., Postacchini, F., Accinni, L., and de Martino, C.,** Morphological, immunochemical and biochemical study of rabbit Achilles tendon at various ages, *J. Bone Jt. Surg.,* 62A, 583, 1980.

42. **Gathercole, L. J. and Keller, A.,** Early development of crimping in rat tail tendon collagen: a polarizing optical and SEM study, *Micron,* 9, 83, 1978.

43. **Niven, H., Baer, E., and Hiltner, A.,** Organization of collagen fibers in rat tail tendon at the optical microscope level, *Collagen Rel. Res.,* 2, 131, 1982.

44. **Serafini-Fracassini, A., Field, J. M., Smith, J. W., and Stephens, W. G. S.,** The ultrastructure and mechanics of elastic ligaments, *J. Ultrastr. Res.,* 58, 244, 1977.

45. **Bergel, D. H.,** The properties of blood vessels, in *Biomechanics, Its Foundations and Objectives,* Fung, Y. C., Perrone, N., and Anliker, M., Eds., Prentice-Hall, Englewood Cliffs, N.J., 1972, chap. 5.

46. **Krane, S. M.,** Collagenases and collagen degradation, *J. Invest. Dermatol.,* 79, 83s, 1982.

47. **Nakagawa, H., Isaji, M., Hayashi, M., and Tsurufuji, S.,** Selective inhibition of collagen breakdown by proteinase inhibitors in granulation tissue in rats, *J. Biochem.,* 89, 1081, 1981.

48. **Tamayo, R. P.,** Morphologic and biochemical aspects of collagen degradation, in *Collagen Degradation and Mammalian Collagenase,* Tsuchiya, M., Tamayo, R. P., Okazaki, I., and Maruyama, K., Eds., Excerpta Medica, Int. Congr. Ser. #601, 1982, 3.

49. **Fisher, W. D., Lyons, H., van der Rest, M., Cooke, T. D. V., and Poole, A. R.,** The stimulation of collagenase from rheumatoid synovium by fragments of cartilage collagen, *Proc. 25th Meet. Orthop. Res. Soc.,* 52, 1979.

50. **Shoshan, S.,** Wound healing, *Int. Rev. Connect. Tiss. Res.,* 9, 1, 1981.

51. **Houck, J. and Chang, C.,** The chemotactic properties of the products of collagenolysis, *Proc. Soc. Exp. Biol. Med.,* 138, 69, 1971.

52. **Klein, L., Dawson, M. H., and Heiple, K. G.,** Turnover of collagen in the adult rat after denervation, *J. Bone Jt. Surg.,* 59A, 1065, 1977.

53. **Michaeli, D., Benjamin, E., Leung, D. Y. K., and Martin, G. R.,** Immunochemical studies on collagen. II. Antigenic differences between guinea pig skin collagen and gelatin, *Immunochemistry,* 8, 1, 1971.

54. **Furthmayr, H. and Timpl, R.,** Immunochemistry of collagens and procollagens, *Int. Rev. Connect. Tiss. Res.,* 7, 61, 1976.

55. **Chvapil, M., Kronenthal, R. L., and van Winkle, W., Jr.,** Medical and surgical applications of collagen, *Int. Rev. Connect. Tiss. Res.,* 6, 1, 1973.

56. **Rosenwasser, L., Bhatnagar, R., and Stobo, J.,** Genetic control of the murine T lymphocyte proliferative response to collagen: analysis of the molecular and cellular contributions to immunogenicity, *J. Immunol.,* 124, 2854, 1980.

57. **Lee, E. H., Langer, F., Halloran, P., Gross, A. E., and Ziv, I.,** The immunology of osteochondral and massive bone allografts, *Proc. 25th Meet. Orthop. Res. Soc.,* 61, 1979.

58. **Bennett, J. G. and Drury, P. J.,** Mechanical comparison of preserved biological vascular grafts, *Proc. 8th Annual Soc. Biomater.,* 103, 1982.

59. **Babin, R. W., Gantz, B. J., and Maynard, J. A.,** Survival of implanted irradiated cartilage, *Otolaryngol. Head Neck Surg.,* 90, 75, 1982.

60. **Geroulanos, S., von Meiss, U., Walter, P., Uhlschmid, G., Turina, M., and Senning, Aa.,** A new vascular prosthesis for small diameter vessel replacement, *Trans. Am. Soc. Artif. Intern. Org.,* 28, 200, 1982.

61. **Loissance, D., Moczar, M., Leandri, J., Bessou, J. P., and David, P. H.,** A new microarterial biograft, *Trans. Am. Soc. Artif. Intern. Organs,* 27, 401, 1981.

62. **Oliver, R. F., Grant, R. A., Hulme, M. J., and Mudie, A.,** Incorporation of stored cell-free dermal collagen allografts into skin wounds: a short term study, *Br. J. Plast. Surg.,* 30, 88, 1977.

63. **Rosenberg, N., Henderson, J., Lord, G. H., Bothwell, J. W., and Gaughran, E. R. L.,** Use of enzyme-treated heterografts as segmental arterial substitutes. V. Influence of processing factors on strength and invasion by host, *Arch. Surg.,* 85, 32, 1962.

64. **Oliver, R. F., Barker, H., Cooke, A., and Grant, R. A.,** Dermal collagen implants, *Biomaterials,* 3, 38, 1982.

65. **Berg, A. and Eckmayer, Z.,** Method for Making Collagen Sponge for Medical and Cosmetic Uses, U.S. Patent 4,320,201, 1982.

66. **Becker, U., Braun, K., and Heimburger, N.,** Collagen Solution, Process for its Manufacture and its Use, U.S. Patent 4,331,766, 1982.

67. **Hochstadt, H. R. and Lieberman, E. R.,** Collagen Article and the Manufacture Thereof, U.S. Patent 2,920,000, 1960.

68. **Reissmann, T. L. and Nichols, J.,** Collagen Article and the Manufacture Thereof, U.S. Patent 2,919,999, 1960.

69. **Price, P. J.,** Preparation and use of rat-tail collagen, *TCA Manual,* 1, 43, 1975.

70. **Gross, J. and Kirk, D.,** The heat precipitation of collagen from neutral salt solutions: some rate-regulating factors, *J. Biol. Chem.,* 233, 355, 1958.

71. **Cassel, J. M., Mandelkern, L., and Roberts, D. E.,** The kinetics of the heat precipitation of collagen, *J. Am. Leather Chem. Assoc.,* 55, 556, 1960.

72. **Liebermann, E. R. and Oneson, I. B.,** Collagen Film, U.S. Patent 3,014,024, 1961.

73. **Schor, S. L.,** Cell proliferation and migration on collagen substrata, *in vitro, J. Cell Sci.,* 41, 159, 1980.

74. **Gelman, R. A., Poppke, D. C., and Piez, K. A.,** Collagen fibril formation *in vitro;* the role of the nonhelical terminal regions, *J. Biol. Chem.,* 254, 11741, 1979.

75. **Danielsen, C. C.,** Thermal stability of reconstituted collagen fibrils; shrinkage characteristics upon *in vitro* maturation, *Mech. Aging Dev.,* 15, 269, 1981.

76. **Suzuki, S., Karube, I., and Satoh, I.,** Electrochemical preparation of enzyme-collagen membrane for enzyme electrode, in *Biomedical Applications of Immobilized Enzymes and Proteins,* Chang, T. M. S., Ed., Plenum Press, New York, 1977, chap. 37.

77. **Knapp, T. R., Luck, E., and Daniels, J. R.,** Behavior of solubilized collagen as a bioimplant, *J. Surg. Res.,* 23, 96, 1977.

78. **Stenzel, K. H., Miyata, T., and Rubin, A. L.,** Collagen as a biomaterial, *Ann. Rev. Biophys. Bioeng.,* 3, 321, 1974.

79. **Miyata, T.,** Collagen Skin Dressing, U.S. Patent 4,294,241, 1981.

80. **Chvapil, M.,** Reconstituted collagen, in *Biology of Collagen,* Viidik, A. and Juust, J., Eds., Academic Press, London, 1980, chap. 22.

81. **Nichols, J.,** Collagen Fibril Matrix Pharmaceuticals, U.S. Patent 3,435,117, 1969.

82. **Stenzel, K. H., Rubin, A. L., Yamayoshi, W., Miyata, T., Suzuki, T., Sohde, T., and Nishizawa, M.,** Optimization of collagen dialysis membranes, *Trans ASAIO,* 17, 293, 1971.

83. **Hanyen, C. K., Nichols, J., and Rich, E. J.,** Laminated Collagen Suture, U.S. Patent 3,518,994, 1970.

84. **Griset, E. J., Reissmann, T. L., and Nichols, J.,** Method of Producing a Collagen Strand, U.S. Patent 3,114,593, 1963.

85. **Chvapil, M.,** Collagen sponge: theory and practice of medical applications, *J. Biomed. Mater. Res.,* 11, 721, 1977.

86. **Iversen, P. L., Partlow, L. M., Stensaas, L. J., and Moatamed, F.,** Characterization of a variety of standard collagen substrates: ultrastructure, uniformity and capacity to bind and promote growth of neurons, *In Vitro,* 17, 540, 1981.

87. **Williams, B. R., Gelman, R. A., Poppke, D. C., and Piez, K. A.,** Collagen fibril formation; optimal *in vitro* conditions and preliminary kinetic results, *J. Biol. Chem.,* 253, 6578, 1978.

88. **Danielsen, C. C.,** Mechanical properties of reconstituted collagen fibrils, *Conn. Tissue Res.,* 9, 51, 1981.

89. **Dunn, G. A. and Ebendal, T.,** Contact guidance on oriented collagen gels, *Exp. Cell Res.,* 111, 475, 1978.

90. **Bianchi, E., Conio, G., Ciferri, A., Puett, E., and Rajagh, L.,** The role of pH, temperature, salt type and salt concentration on the stability of the crystalline, helical and randomly coiled forms of collagen, *J. Biol. Chem.,* 262, 1361, 1967.

91. **Gelman, R. A., Williams, B. R., and Piez, K. A.,** Collagen fibril formation; evidence for a multistep process, *J. Biol. Chem.,* 254, 180, 1979.

92. **Wallace, D. G. and Thompson, A.,** Description of collagen fibril formation by a theory of polymer crystallization, *Biopolymers,* 22, 1793, 1983.

93. **Weiss, B. A.,** Stress relaxation in aged reconstituted type I collagen suspension, *Trans. 9th Annu. Soc. Biomater.,* 6, 51, 1983.

94. **Masurovsky, E. B. and Peterson, E. R.,** A procedure for preparing photo-gelled collagen substrates containing Aclar reticles for tissue culture, *TCA Manual,* 1, 107, 1975.

95. **Marino, A. A., Spadaro, J. A., Fukada, E., Kahn, L. D., and Becker, R. O.,** Piezoelectricity in collagen films, *Calcif. Tissue Int.*, 31, 257, 1980.

96. **Weiss, B. A.** Unpublished data, Collagen Corporation, 1983.

97. **Oliver, R. F., Grant, R. A., Cox, R. W., and Cooke, A.,** Effect of aldehyde cross-linking on human dermal collagen implants in the rat, *Br. J. Exp. Pathol.*, 61, 544, 1980.

98. **Huang, C. and Yannas, I. V.,** Mechanochemical studies of enzymatic degradation of insoluble collagen fibers, *J. Biomed. Mater. Res. Symp.*, 8, 137, 1977.

99. **Chvapil, M., Owen, J. A., and Clark, D. S.,** Effect of collagen crosslinking on the rate of resorption of implanted collagen tubing in rabbits, *J. Biomed. Mater. Res.*, 11, 297, 1977.

100. **Speer, D. P., Chvapil, M., Volz, R. G., and Holmes, M. D.,** Enhancement of healing in osteochondral defects by collagen sponge implants, *Clin. Orthop. Rel. Res.*, 144, 326, 1979.

101. **Speer, D. P., Chvapil, M., Eskelson, C. D., and Ultreich,** Biological effects of residual glutaraldehyde in glutaraldehyde-tanned collagen biomaterials, *J. Biomed. Mater. Res.*, 14, 753, 1980.

102. **Xenotech Laboratories, Inc.,** Tendon Xenograft bioprosthesis Model CLR, product literature, 1983.

103. **Shimizu, Y., Matsunobe, M., Miyamoto, Y., Watanabe, S., Kato, H., Matsumoto, M., Teramatsu, T., Hino, T., and Okamura, S.,** Study on composite of collagen and synthetic polymer, *Proc. 2nd ISAO*, 197, 1979.

104. **Lawrence, C. A. and Block, S. S., Eds.,** *Disinfection, Sterilization and Preservation*, Lea & Febiger, Philadelphia, 1968.

105. **Hawrot, E.,** Cultured sympathetic neurons: effects of cell-derived and synthetic substrata on survival and development, *Dev. Biol.*, 74, 136, 1980.

106. **Vroom, D. A.,** Electron-beam sterilization, an alternative to gamma irradiation, *Med. Dev. Diag. Ind.*, 2, 35, 1980.

107. **Nevins, A. J.,** Endodontic Composition and Method, U.S. Patent 3,968,567, 1976.

108. **Chvapil, M.,** personal communication, 1983.

109. **Stonehill, A. A., Krop, S., and Borick, P. M.,** Buffered glutaraldehyde, a new chemical sterilizing solution, *Am. J. Hosp. Pharm.*, 20, 458, 1963.

110. **Prolo, D. J., Pedrotti, P. W., and White, D. H.,** Ethylene oxide sterilization of bone, dura mater and fascia lata for human transplantation, *Neurosurgery*, 6, 529, 1980.

111. **Schnell, J., Plenio, H. G., Pauli, W., Braun, B., and Korb, G.,** The influence of ionizing radiation on various collagen-containing medical products, *Radiosterilization of Medical Products*, Paper SM92/13 International Atomic Energy Agency, Vienna, 1967, 145.

112. **Sabelman, E. E., Armstrong, R., and Seyedin, S.,** Osteoinductive activity of xenogeneic demineralized bone, *Proc. 9th Annu. Meet. Soc. Biomater.*, 6, 22, 1983.

113. **Griffiths, R. W. and Shakespeare, P. G.,** Human dermal collagen allografts: a three year histological study, *Br. J. Plas. Surg.*, 35, 519, 1982.

114. **Lee, H. and Neville, K.,** *Handbook of Biomedical Plastics*, Pasadena Technology Press, Pasadena, Calif., 1971.

115. **Marion, L., Haugen, E., and Mjoer, I. A.,** Methodological assessments of subcutaneous implantation techniques, *J. Biomed. Mater. Res.*, 14, 343, 1980.

116. **Murphy, W. M.,** Tissue reaction of rats and guinea pigs to Co-Cr implants with different surface finishes, *Br. J. Exp. Pathol.*, 52, 353, 1971.

117. **Misiek, D. J., Kent, J. N., Carr, R. F., and Saal, C. J.,** The inflammatory response to different shaped hydroxylapatite particles implanted in soft tissues, *Proc. 9th Annu. Meet. Soc. Biomater.*, 6, 23, 1983.

118. **Matlaga, B. F., Yasenchak, L. P., and Salthouse, T. N.,** Tissue response to implanted polymers: the significance of sample shape, *J. Biomed. Mater. Res.*, 10, 391, 1976.

119. **Dunn, G. A. and Heath, J. P.,** A new hypothesis of contact guidance in tissue cells, *Exp. Cell Res.*, 101, 1, 1976.

120. **Picha, G. J.,** Tissue response to peritoneal implants, NASA CR-159817, 1980.

121. **Gozna, E. R. and Marble, A. E.,** Optimal design parameters and performance of arterial grafts, in *Biomechanics of Medical Devices*, Ghista, D. N., Ed., Marcel Dekker, New York, 1981, chap. 7.

122. **Carson, S., Demling, R., Esquivel, C., Talken, L., and Tillman, P.,** Testing and treatment of arterial graft thrombosis, *Am. J. Surg.*, 142, 137, 1981.

123. **Christie, B. A., Ketharanathan, V., and Perloff, L. J.,** Minute vascular replacements, *Arch. Surg.*, 117, 1290, 1982.

124. **Field, P. L., Milne, P. Y., Ketharanathan, V., and Macleish,** The Australian biosynthetic vascular graft — ovine collagen and dacron mesh — clinical results and performance as a small vessel substitute implanted in humans, *Proc. 8th Annu. Meet. Soc. Biomater.*, 105, 1982.

125. **Krajicek, M., Zastava, V., and Chvapil, M.,** Collagen-fabric vascular prostheses; biological and morphological experience, *J. Surg. Res.*, 4, 290, 1964.

126. **Noishiki, Y. and Miyata, T.,** Initial healing process of succinylated collagen tube as an antithrombogenic cardiovascular graft, *Proc. 8th Annu. Meet. Soc. Biomater.*, 102, 1982.

127. **Krippaehna, W. W.,** Reaction of connective tissue to collagen-Dacron prosthesis, *Am. J. Surg.,* 110, 186, 1965.

128. **Noishiki, Y. and Miyata, T.,** A method to give an antithrombogenicity to biological materials, *Proc. 9th Annu. Meet. Soc. Biomater.,* 17, 1983.

129. **Raghunath, K., Biswas, G., Rao, K. P., Joseph, K. T., and Chvapil, M.,** Some characteristics of collagen-heparin complex, *J. Biomed. Mater. Res.,* 17, 613, 1983.

130. **Silver, M. M., Pollock, J., Silver, M. D., Williams, W. G., and Trusler, G. A.,** Calcification in porcine xenogrˈft valves in children, *Am. J. Cardiol.,* 45, 685, 1980.

131. **Levy, ʀ. J., Schoen, F. J., Howard, S. L., and Oshry, L.,** Mechanisms of glutaraldehyde preserved porcine aortic valve calcification, *Proc. 9th Annu. Meet. Soc. Biomater.,* 98, 1983.

132. **Harasaki, H., Fields, A., McMahon, J., and Nose, Y.,** Tissue valve calcification, *Proc. 9th Annu. Meet. Soc. Biomater.,* 97, 1983.

133. **Lentz, D. J., Pollock, E. M., Olsen, D. B., and Andrews, E. J.,** Prevention of intrinsic calcification in porcine and bovine xenograft materials, *Trans. ASAIO,* 28, 494, 1982.

134. **Kronick, P. and Jimenez, S. A.,** The size of collagen fibrils that stimulate platelet aggregation in human plasma, *Biochem. J.,* 186, 5, 1980.

135. **Zucker, W. H. and Mason, R. G.,** Ultrastructural aspects of interactions of platelets with microcrystalline collagen, *Am. J. Pathol.,* 82, 129, 1976.

136. **Colen, L. B. and Mathes, S. J.,** The use of microcrystalline collagen in microsurgery and its effect on anastomotic patency, *Ann. Plas. Surg.,* 9, 471, 1982.

137. **Saroff, S. A., Chasens, A. I., Eisen, S. F., and Levey, S. H.,** Free soft tissue autografts—hemostasis and protection of the palatal donor site with microfibrillar collagen preparation, *J. Periodont.,* 53, 425, 1982.

138. **Silverstein, M. E. and Chvapil, M.,** Experimental and clinical experiences with collagen fleece as a hemostatic agent, *J. Trauma,* 21, 388, 1981.

139. **Silverstein, M. E., Keown, K., Owen, J. A., and Chvapil, M.,** Collagen fibers as a fleece hemostatic agent, *J. Trauma,* 20, 688, 1980.

140. **Coln, D., Horton, J., Ogden, M. E., and Buja, L. N.,** Evaluation of hemostatic agents in experimental splenic lacerations, *Am. J. Surg.,* 145, 256, 1983.

141. **Sawyer, P. N., Stanczewski, B., Fitzgerald, J., Sivasubramanian, P. P., Mistry, F., and Landi, J.,** A new effective intraoperative hemostat — superstat, *Trans. ASAIO,* 26, 507, 1980.

142. **Senn, N.,** On the healing of aseptic bone cavities by implantation of antiseptic decalcified bone, *Am. J. Med. Sci.,* 98, 219, 1889.

143. **Urist, M. R.,** The substratum for bone morphogenesis, *Dev. Biol.,* 4(Suppl.), 125, 1970.

144. **Sampath, T. K. and Reddi, A. H.,** Dissociative extraction and reconstitution of extracellular matrix components involved in local bone differentiation, *Proc. Natl. Acad. Sci. U.S.A.,* 78, 7599, 1981.

145. **Libin, B. M., Ward, H. L., and Fishman, L.,** Decalcified, lyophilized bone allografts for use in human periodontal defects, *J. Periodontol.,* 56, 51, 1975.

146. **Mulliken, J. B. and Glowacki, J.,** Induced osteogenesis for repair and construction in the craniofacial region, *Plas. Recon. Surg.,* 65, 553, 1980.

147. **Krukowski, M. and Kahn, A. J.,** Inductive specificity of mineralized bone matrix in ectopic osteoclast differentiation, *Calcif. Tissue Int.,* 34, 474, 1982.

148. **Chambers, T. J.,** Phagocytic recognition of bone by macrophages, *J. Pathol.,* 135, 1981.

149. **Yakagi, Y., Kuboki, Y., and Sasaki, S.,** Detection of collagen degradation products from subcutaneously implanted organic bone matrix, *Calcif. Tissue Int.,* 28, 253, 1979.

150. **Anderson, H. C. and Griner, S. A.,** Cartilage induction *in vitro, Dev. Biol.,* 60, 351, 1977.

151. **DeVore, D. T.,** Collagen xenografts for bone replacement: the effects of aldehyde-induced cross-linking on degradation rate, *Oral Surg.,* 43, 677, 686, 1977.

152. **Gross, B. P., Nevins, A. J., and Laporta, R.,** Bone induction potential of mineralized collagen gel xenografts, *Oral Surg.,* 49, 21, 1980.

153. **Miyata, T., Akiyama, T., and Furose, M.,** Artificial Bone, U.S. Patent 4,314,380, 1982.

154. **Joos, V., Ochs, G., and Ries, P.,** Influence of collagen fleece on bone regeneration, *Biomaterials,* 1, 23, 1980.

155. **Cucin, R. L., Goulian, D., Jr., Stenzel, K. H., and Rubin, A. L.,** The effect of reconstituted collagen gels on the healing of experimental bony defects: a preliminary report, *J. Surg. Res.,* 12, 318, 1972.

156. **Wilkinson, H. A., Baker, S., and Rosenfeld, S.,** Gelfoam paste in experimental laminectomy and cranial trephination, *J. Neurosurg.,* 54, 664, 1981.

157. **Rosenberg, L. C.,** Biological basis for the imperfect repair of articular cartilage following injury, *Int. Symp. Tissue Repair,* 26, 1983.

158. **Narang, R. and Wells, H.,** Decalcified allogeneic bone matrix implantation in joint spaces of rats, *Oral Surg.,* 35, 730, 1973.

159. **Bimstein, E. and Shoshan, S.,** Enhanced healing of tooth-pulp wounds in the dog by enriched collagen solution as a capping agent, *Arch. Oral Biol.,* 26, 97, 1981.
160. **Parsons, J. R., Alexander, H., and Weiss, A. B.,** Soft tissue repair and replacement with carbon fiber/absorbable polymer composites, *Int. Symp. Tissue Repair,* 83, 1983.
161. **Chvapil, M.,** Considerations on manufacturing principles of a synthetic burn dressing: a review, *J. Biomed. Mater. Res.,* 16, 245, 1982.
162. **Yannas, I. V. and Burke, J. F.,** Design of an artificial skin. I. Basic design principles, *J. Biomed. Mater. Res.,* 14, 65, 1980.
163. **Yannas, I. V., Burke, J. F., Gordon, P. L., Huang, C., and Rubenstein, R. H.,** Design of an artificial skin. II. Control of chemical composition, *J. Biomed. Mater. Res.,* 14, 107, 1980.
164. **Yannas, I. V., Burke, J. F., Orgill, D. P., and Skrabut, E. M.,** Wound tissue can utilize a polymeric template to synthesize a functional extension of skin, *Science,* 215, 174, 1982.
165. **Bell, E., Ehrlich, H. P., Buttle, D. J., and Nakatsuji, T.,** Living tissue formed in vitro and accepted as skin-equivalent tissue of full thickness, *Science,* 211, 1052, 1981.
166. **Morykwas, M. J., Thornton, J. W., Feller, I., and Bartlett, R. H.,** Surface charge of synthetic and biological skin substitutes: effects on adherence, *Proc. 9th Annu. Meet. Soc. Biomater.,* 5, 1983.
167. Collagen Corporation, Zyderm collagen implant, Physician package insert, code 7004, 1982.
168. Collagen Corporation, Long-term efficacy and maintenance, code 7016, 1982.
169. Collagen Corporation, Zyderm collagen implant, Summary of clinical investigation, code 7047, 1982.
170. **Kamer, F. and Hunter, D.,** Use of injectable collagen for cosmetic lines of the face: preliminary report, *Laryngoscope,* 93, 950, 1983.
171. **Barr, R. J., King, D. F., McDonald, R. M., and Bartlow, G. A.,** Necrobiotic granulomas associated with bovine collagen test site injections, *J. Am. Acad. Dermatol.,* 6, 867, 1982.
172. **Hanke, C. W. and Robinson, J. K.,** Injectable collagen implants, *Arch. Dermatol.,* 119, 533, 1983.
173. **Armstrong, R., Cooperman, L. S., and Parkinson, T. A.,** Injectable collagen for soft tissue augmentation, National Institutes of Health Consensus Panel Meeting, in *Contemporary Biomaterials: Material and Host Response, Clinical Applications, New Technology, and Legal Aspects,* Boretos, J. W. and Eden, M., Eds., Noyes, Park Ridge, N.J., 1984.
174. **Driller, J. and Penn, R. D.,** Selective vascular occlusions: a material evaluation, *Proc. 29th Ann. Conf. Eng. Med. Biol.,* Paper 23.3, 170, 1976.
175. **Marzin, L. and Rouveix, B.,** An evaluation of collagen gel in chronic leg ulcers, *Schweiz. Rundschau. Med. (Praxis),* 71, 1373, 1982.
176. **Klibanov, A. M.,** Immobilized enzymes and cells as practical catalysts, *Science,* 219, 722, 1983.
177. **Arrio-Dupont, M. and Coulet, P. R.,** Role of diffusion of substrates on the apparent behaviour of immobilized malate dehydrogenase, *FEBS Lett.,* 143, 279, 1982.
178. **Lee, K. H-K., Coulet, P. R., and Gautheron, D. C.,** Grafting of enzymes on collagen films using Woodward's reagent ''K'' and a water-soluble carbodiimide derivative, *Biochimie,* 58, 489, 1976.
179. **Bernath, F. R., Olanoff, L. S., and Vieth, W. R.,** Therapeutic perspectives of enzyme reactors, in *Biomedical Applications of Immobilized Enzymes and Proteins,* Chang, T. M. S., Ed., 1977, chap. 23.
180. **Gondo, S. and Hoya, H.,** Solubilized collagen fibril as a supporting material for enzyme immobilization, *Biotech. Bioeng.,* 20, 2007, 1978.
181. **Wang, S. S. and Vieth, W. R.,** Collagen-enzyme complex membranes and their performance in biocatalytic modules, *Biotech. Bioeng.,* 15, 93, 1973.
182. **Karube, I., Yugeta, Y., and Suzuki, S.,** Electric field control of lipase membrane activity, *Biotech. Bioeng.,* 19, 1493, 1977.
183. **Watanabe, S., Shimizu, Y., Teramatsu, T., Murachi, T., and Hino, T.,** The *in vitro* and *in vivo* behavior of urokinase immobilized onto collagen-synthetic polymer composite material, *J. Biomed. Mater. Res.,* 15, 553, 1981.
184. **Dorr, R. T., Surwitt, E. A., Droegemueller, W., Alberts, D. S., Meyskens, F. L., and Chvapil, M.,** *In vitro* retinoid binding and release from a collagen sponge material in a simulated intravaginal environment, *J. Biomed. Mater. Res.,* 16, 839, 1982.
185. **Rubin, A. L., Stenzel, K. H., Miyata, T., White, M. J., and Dunn, M.,** Collagen as a vehicle for drug delivery, *J. Clin. Pharmacol.,* 13, 309, 1973.
186. **Slatter, D. H., Costa, N. D., and Edwards, M. E.,** Ocular inserts for application of drugs to bovine eyes — *in vivo* and *in vitro* studies on the release of gentamicin from collagen inserts, *Austr. Vet. J.,* 59, 4, 1982.
187. **Abbenhaus, J. I. and Hemenway, W. G.,** Bovine collagen as a tympanic membrane graft in dogs, *Surg. Forum.,* 18, 490, 1967.
188. **Bellucci, R. J. and Wolff, D.,** Experimental stapedectomy with collagen sponge implant, *Laryngoscope,* 74, 668, 1964.
189. **Liston, S. L.,** Microfibrillar collagen in the oval window, *Otolaryngol. Head Neck Surg.,* 90, 844, 1982.

190. **Chole, R. A.,** Ossicular replacement with self-stabilizing presculptured homologous cartilage, *Arch. Otolarygol.,* 108, 560, 1982.

191. **Kline, D. G. and Hayes, G. J.,** The use of a resorbable wrapper for peripheral-nerve repair, *J. Neurosurg.,* 21, 737, 1964.

192. **Rosen, J. M., Kaplan, E. N., Jewett, D. L., and Daniels, J. R.,** Fascicular sutureless and suture repair of the peripheral nerves; a comparison study in laboratory animals, *Orthop. Rev.,* 8, 85, 1979.

193. **Ehrmann, R. L. and Gey, G. O.,** The growth of cells on a transparent gel of reconstituted rat-tail collagen, *J. Natl. Cancer Inst.,* 16, 1375, 1956.

194. **Dodd, N. J. F., Schor, S. L., and Rushton, G.,** The effects of a collagenous extra-cellular matrix on fibroblast membrane organization, *Exp. Cell Res.,* 141, 421, 1982.

195. **Richards, J., Guzman, R., Konrad, M., Yang, J., and Nandi, S.,** Growth of mouse mammary gland end buds cultured in a collagen gel matrix, *Exp. Cell Res.,* 141, 433, 1982.

196. **Talley, D. J., Roy, W. A., and Li, J. J.,** Behavior of primary and serially transplanted estrogen-dependent renal carcinoma cells in monolayer and in collagen gel culture, *In Vitro,* 18, 149, 1982.

197. **Schor, A. M., Schor, S. L., and Allen, T. D.,** Effects of culture conditions on the proliferation, morphology and migration of bovine aortic endothelial cells, *J. Cell Sci.,* 62, 267, 1983.

198. **Brown, A. F.,** Neutrophil granulocytes: adhesion and locomotion on collagen substrata and in collagen matrices, *J. Cell Sci.,* 58, 455, 1982.

199. **Gibson, G. J., Schor, S. L., and Grant, M. E.,** Effects of matrix macromolecules on chondrocyte gene expression: synthesis of a low molecular weight collagen species by cells cultured within collagen gels, *J. Cell Biol.,* 93, 767, 1982.

200. **Schor, S. L. and Court, J.,** Different mechanisms in the attachment of cells to native and denatured collagen, 38, 267, 1979.

201. **Grinnell, F. and Bennett, M. H.,** Fibroblast adhesion on collagen substrata in the presence and absence of plasma fibronectin, *J. Cell Sci.,* 48, 19, 1981.

202. **Richards, J., Pasco, D., Yang, J., Guzman, R., and Nandi, S.,** Comparison of the growth of normal and neoplastic mouse mammary cells on plastic, on collagen gels and in collagen gels, *Exp. Cell Res.,* 146, 1, 1983.

203. **Linsenmayer, T. F., Gibney, E., Toole, B. P., and Gross, J.,** Cellular adhesion to collagen, *Exp. Cell Res.,* 116, 470, 1978.

204. **Elsdale, T. and Bard, J.,** Collagen substrata for studies on cell behavior, *J. Cell Biol.,* 54, 626, 1972.

205. **Kleinman, H. K., McGoodwin, E. B., Rennard, S. I., and Martin, G. R.,** Preparation of collagen substrates for cell attachment: Effect of collagen concentration and phosphate buffer, *Anal. Biochem.,* 94, 308, 1979.

206. **Haston, W. S., Shields, J. M., and Wilkinson, P. C.,** Lymphocyte locomotion and attachment on two-dimensional surfaces and in three-dimensional matrices, *J. Cell. Biol.,* 92, 747, 1982.

207. **Jacobson, B. S. and Ryan, U. S.,** Growth of endothelial and HeLa cells on a new multipurpose microcarrier that is positive, negative or collagen coated. *Tissue Cell,* 14, 69, 1982.

208. **Aplin, J. D., Bardsley, W. G., and Niven, V. M.,** Kinetic analysis of cell spreading. II. Substratum adhesion requirements of amniotic epithelial (FL) cells, *J. Cell Sci.,* 61, 375, 1983.

209. **Bullaro, J. C.,** Monolayer cell cultures of embryonic chick skeletal muscle, *J. Tiss. Cult. Meth.,* 6, 91, 1980.

210. **Michalopoulos, G. and Pitot, H. C.,** Primary culture of parenchymal liver cells on collagen membranes, *Exp. Cell. Res.,* 94, 70, 1975.

211. **Daniels, E.,** Three-dimensional analysis of hemopoietic cellular interactions *in vitro* using Gelfoam sponge matrices, *J. Tiss. Cult. Meth.,* 6, 65, 1980.

212. **Stratton, C. J., Douglas, W. H. J., and McAteer, J. A.,** The surfactant system of human fetal lung organotypic cultures: ultrastructural preservation by a lipid-carbohydrate retention method, *Anat. Rec.,* 192, 481, 1978.

213. **McAteer, J. A., Cavanaugh, T. J., and Evan, A. P.,** Submersion culture of the intact fetal lung, *In Vitro,* 19, 210, 1983.

214. **Freeman, A. E., Yoshida, Y., Hilborn, V., and Carey, S. L.,** Culturing epithelial cell types on a pigskin substrate, *TCA Manual,* 5, 1181, 1979.

215. **Leighton, J.,** Collagen-coated cellulose sponge, 1 in *Tissue Culture; Methods and Applications,* Kruse, P. F., Jr. and Patterson, M. K., Jr., Eds., Academic Press, New York, 1973, chap. 1.

216. **Gabridge, M. G. and Glass, M. F.,** Collagen as an attachment factor for cell monolayers on gas-permeable Teflon membranes, *In Vitro,* 19 (Abstr.), 259, 1983.

217. **Aplin, J. D. and Hughes, R. C.,** Protein-derivatised glass coverslips for the study of cell-to-substratum adhesion, *Anal. Biochem.,* 113, 144, 1981.

218. **Lash, J. W. and Vasan, N. S.,** Tissue interactions and extracellular matrix components, in *Cell and Tissue Interactions,* Lash, J. W. and Burger, M. M., Eds., Raven Press, New York, 1977, 101.

219. **Goldberg, B. D. and Burgeson, R. E.,** Binding of soluble Type I collagen to fibroblasts: specificities for native collagen types, triple helical structure, telopeptides, propeptides and cyanogen bromide-derived peptides, *J. Cell. Biol.,* 95, 752, 1982.

220. **Schor, S. L., Schor, A. M., and Bazill, G. W.,** The effects of fibronectin on the migration of human foreskin fibroblasts and Syrian hamster melanoma cells into three-dimensional gels of native collagen fibres, *J. Cell Sci.,* 48, 301, 1981.

221. **Murray, J. C., Stingl, G., Kleinman, H. K., Martin, G. R., and Katz, S. I.,** Epidermal cells adhere preferentially to Type IV (basement membrane) collagen, *J. Cell Biol.,* 80, 197, 1979.

222. **Grinnell, F.,** Migration of human neutrophils in hydrated collagen lattices, *J. Cell Sci.,* 58, 95, 1982.

223. **Grobstein, C.,** Developmental role of intercellular matrix: retrospective and prospective, in *Extracellular matrix influences on gene expression,* Slavkin, H. C. and Greulich, R. C., Eds., Academic Press, New York, 1975, 9.

224. **Linsenmayer, T. F. and Toole, B. P.,** Biosynthesis of different collagens and glycosaminoglycans during limb development, *Birth Defects: Orig. Artic. Ser.,* 13, 19, 1977.

225. **Caplan, A. I., Fiszman, M. Y., and Eppenberger, H. M.,** Molecular and cell isoforms during development, *Science,* 221, 921, 1983.

226. **Bellows, C. G., Melcher, A. H., Bhargava, U., and Aubin, J. E.,** Fibroblasts contracting three-dimensional collagen gels exhibit ultrastructure consistent with either contraction or protein secretion, *J. Ultrastr. Res.,* 78, 178, 1982.

227. **Lee, E. Y-H., Parry, G., and Bissell, M. J.,** Modulation of secreted proteins of mouse mammary epithelial cells by the collagenous substrata, *J. Cell Biol.,* 98, 146, 1983.

228. **Wicha, M. S., Lowrie, G., Kohn, E., Bagavandoss, P., and Mahn, T.,** Extracellular matrix promotes mammary epithelial growth and differentiation *in vitro, Proc. Natl. Acad. Sci. U.S.A.,* 79, 3213, 1982.

229. **Ormerod, E. J. and Rudland, P. S.,** Mammary gland morphogenesis *in vitro:* formation of branched tubules in collagen gels by a cloned rat mammary cell line, *Dev. Biol.,* 91, 360, 1982.

230. **Dessau, W., Vertel, B. M., von der Mark, H., and von der Mark, K.,** Extracellular matrix formation by chondrocytes in monolayer culture, *J. Cell Biol.,* 90, 78, 1981.

231. **Benya, P. D. and Shaffer, J. D.,** Dedifferentiated chondrocytes reexpress the differentiated collagen phenotype when cultured in agarose gels, *Cell,* 30, 215, 1982.

232. **Solursh, M., Linsenmayer, T. F., and Jensen, K. L.,** Chondrogenesis from single limb mesenchyme cells, *Dev. Biol.,* 94, 259, 1982.

233. **Schor, A. M., Schor, S. L., Weiss, J. B., Brown, R. A., Kumar, S., and Phillips, P.,** A requirement for native collagen in the mitogenic effect of a low molecular weight angiogenic factor on endothelial cells in culture, *Br. J. Cancer,* 41, 790, 1980.

234. **Montesano, R., Mouron, P., Amherdt, M., and Orci, L.,** Collagen matrix promotes reorganization of pancreatic endocrine cell monolayers into islet-like organoids, *J. Cell Biol.,* 97, 935, 1983.

235. **Bunge, R. P. and Bunge, M. B.,** Evidence that contact with connective tissue matrix is required for normal interaction between Schwann cells and nerve fibers, *J. Cell Biol.,* 78, 943, 1978.

236. **Tomasek, J. J., Hay, E. D., and Fujiwara, K.,** Collagen modulates cell shape and cytoskeleton of embryonic corneal and fibroma fibroblasts: distribution of actin, alpha-actinin and myosin, *Dev. Biol.,* 92, 107, 1982.

237. **Bassett, C. A. L.,** Biologic significance of piezoelectricity, *Calc. Tissue Res.,* 1, 252, 1968.

238. **Gjelsvik, A.,** Bone remodeling and piezoelectricity. I, *J. Biomech.,* 6, 69, 1973.

239. **Gjelsvik, A.,** Bone remodeling and piezoelectricity. II, *J. Biomech.,* 6, 187, 1973.

240. **Athenstaedt, H., Claussen, H., and Schaper, D.,** Epidermis of human skin: pyroelectric and piezoelectric sensor layer, *Science,* 216, 1018, 1982.

241. **Williams, W. S. and Breger, L.,** Piezoelectricity in tendon and bone, *J. Biomech.,* 8, 407, 1975.

242. **Fukada, E., Ueda, H., and Rinaldi, R.,** Piezoelectric and related properties of hydrated collagen, *Biophys. J.,* 16, 911, 1976.

243. **Pollack, S. R., Korostoff, E., Sternberg, M. E., and Koh, J.,** Stress-generated potentials in bone: effects of collagen modifications, *J. Biomed. Mater. Res.,* 11, 677, 1977.

244. **Eriksson, C. and Jones, S.,** Bone mineral and surface charge, *Clin. Orthop. Rel. Res.,* 128, 351, 1977.

245. **Stump, R. F. and Robinson, K. R.,** *Xenopus* neural crest cell migration in an applied electrical field, *J. Cell Biol.,* 97, 1226, 1983.

246. **Tainer, J A., Getzoff, E. D., Alexander, H., Houghten, R. A., Olson, A. J., Lerner, R. A., and Hendrickson, W. A.,** The reactivity of anti-peptide antibodies as a function of the atomic mobility of sites in a protein, *Nature (London),* 312, 127, 1984.

Chapter 4

# POLY-α-AMINO ACIDS AS BIOMEDICAL POLYMERS

**James M. Anderson, Karen L. Spilizewski, and Anne Hiltner**

## TABLE OF CONTENTS

## I. INTRODUCTION

In recent years, the search for new biomedical materials has greatly intensified. Unfortunately, this search has been hindered by the stringent requirements that a given biomaterial be mechanically and biologically durable, and also compatible with blood and/or tissue. In an attempt to overcome the many-faceted problem of biocompatibility, efforts have been directed toward the use of natural-like materials. The main concept behind this approach has been the utilization of the same or similar building blocks, i.e., chemical units, as found in the body.

As proteins composed of α-amino acids are the most abundant macromolecules in humans, many investigators have turned toward studying the poly-α-amino acids as potential biomedical polymers. The foundation for this approach lies in the fact that a massive amount of literature is available on the successful use of synthetic poly-α-amino acids as models in structural, biological, immunological, chemotherapeutic, and enzymological studies. Numerous monographs are available for appreciating these aspects of poly-α-amino acids and it is not the purpose of this paper to review these studies.[1-6] Rather, this paper will be directed toward a review of these studies which have directly dealt with poly-α-amino acids as potential biomedical polymers. Table 1 provides a list of the significant studies on these materials with a view towards their application as biomedical materials. As can be seen, these studies range from an examination of polymer properties to the blood and tissue compatibility of these materials. A small section has been included on the immunogenicity of poly-α-amino acids as this has been a concern of investigators working with these materials.

Poly-α-amino acids of various types have been investigated for their use in specific biomedical applications. As can be seen from Table 2, the potential application of these materials in various applications is extremely broad.

## II. POLYMER PROPERTIES

### A. Structure

In general, poly-α-amino acids may be considered to be a type of nylon 1 polymers. Figure 1 shows the general structural formula where the R group attached to the asymmetric carbon in the molecule determines the specific type of amino acid. As an asymmetric carbon is present in the amino acid and therefore in the monomer units in the polymer, both L and D configurations are possible. Proteins are comprised of L-amino acids and the vast majority of studies carried out on polyamino acids as biomedical polymers have been with polymers using L-amino acids. For the sake of simplicity, it will be assumed that the amino acids discussed in this paper are of the L configuration unless specifically denoted.

Table 3 contains a list of α-amino acids commonly used in poly-α-amino acids for biomedical application. The γ-benzyl ester of glutamic acid, the β-benzyl ester of aspartic acid, and the ε-N-carbobenzoxy derivative of lysine have been included as they are important in the synthesis of polymers containing these amino acids. The derivatives do not occur in nature but are necessary in the synthesis of these amino acids containing glutamic acid, aspartic acid, or lysine. These derivatives function as blocking groups so that the respective carboxyl or amino functions do not participate in side reactions in the preparation of the polymer. As will be seen, amino acids have been used to prepare homopolymers, random copolymers, and block copolymers. Graft copolymers have also been prepared but have not found wide use in studies directed towards the application of polyamino acids as biomedical polymers.

While poly-α-amino acids can be synthesized by a number of solution and solid state techniques, the quantities necessary for characterization by many techniques and testing by in vitro and in vivo methods requires relatively large quantities. Synthesis of these large

**Table 1**
**STUDIES RELATED TO THE APPLICATION OF**
**POLY-α-AMINO ACIDS AS BIOMATERIALS**

|     |     |     |
| --- | --- | --- |
| I.  | Polymer properties | |
|     | A. | Structure |
|     | B. | Surface properties |
|     | C. | Morphology |
|     | D. | Mechanical properties |
|     | E. | Permeability and diffusion |
| II. | Blood compatibility | |
|     | A. | Coagulation |
|     | B. | Platelet interaction |
|     | C. | Thrombosis |
| III.| Tissue response | |
|     | A. | Inflammatory reaction |
|     | B. | Fibrous capsule formation |
|     | C. | Cell attachment |
| IV. | Biodegradation | |
| V.  | Antigenicity of poly-α-amino acids | |

**Table 2**
**BIOMEDICAL**
**APPLICATIONS**

Sutures
Artificial skin
Hemodialysis membranes
Sustained drug release systems

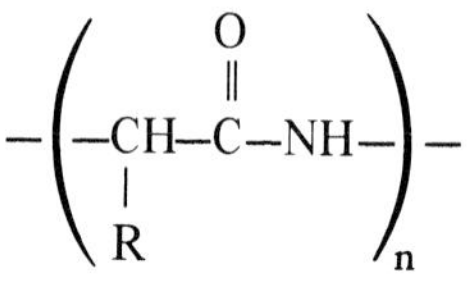

$$-\left(-\mathrm{CH}-\overset{\overset{\textstyle O}{\|}}{\mathrm{C}}-\mathrm{NH}-\right)_{n}-$$

$$\qquad\quad\mathrm{R}$$

Amino acid monomer structure

FIGURE 1

quantities, usually greater than 10 g, is best carried out using polymerization of the corresponding *N*-carboxy-α-amino acid anhydrides (NCAs) in suitable solvents.[1-6] This method does not permit the synthesis of sequential poly-α-amino acids but random, block, and graft copolymers can be easily prepared. Tables 4 and 5 list the various types of random and block copolymers which have been prepared for biomedical polymer investigations and applications.

The synthesis of random and block copolypeptides have generally used the concept of the development of structure/property relationships in polymeric materials. That is, two monomers are utilized in the synthesis of a copolymer with each monomer contributing its specific properties to the overall performance of the copolymer. In this fashion, polymeric materials are created which exhibit a broader range of desirable properties. Considering the diversified nature of the properties necessary for biomedical application of a material, this approach has validity and has been used extensively in providing polyamino acids with a wide variety of properties.

## Table 3
### AMINO ACIDS COMMONLY USED IN POLY-α-AMINO ACIDS FOR BIOMEDICAL POLYMER APPLICATION

| Amino acid | R |
|---|---|
| γ-Benzyl glutamate | $-CH_2-CH_2-COOC_6H_5$ |
| Glutamic acid | $-CH_2-CH_2-COOH$ |
| $N^5$-(2-Hydroxyethyl) glutamine | $-CH_2-CH_2-CONHCH_2CH_2OH$ |
| ε-Carbobenzoxy-lysine | $-CH_2-CH_2-CH_2-NHCOOCH_2C_6H_5$ |
| Lysine | $-CH_2-CH_2-CH_2-CH_2-CH_2-NH_2$ |
| Leucine | $-CH_2-CH(CH_3)_2$ |
| Alanine | $-CH_3$ |
| Phenylalanine | $-CH_2C_6H_5$ |
| Valine | $-CH(CH_3)_2$ |
| β-Benzyl aspartate | $-CH_2-COOC_6H_5$ |
| Aspartic acid | $-CH_2-COOH$ |
| Methionine | $-CH_2-S-CH_3$ |

## Table 4
### RANDOM COPOLYMERS: $(A)_x(B)_y$

| A component | B component |
|---|---|
| γ-Benzyl glutamate | Leucine |
| γ-Benzyl glutamate | Valine |
| γ-Benzyl glutamate | Phenylalanine |
| γ-Benzyl glutamate | Alanine |
| γ-Benzyl glutamate | γ-Methyl glutamate |
| γ-Methyl glutamate | Leucine |
| γ-Benzyl glutamate | Butadiene |
| β-Benzyl aspartate | γ-Benzyl glutamate |
| β-Benzyl aspartate | ε-Carbobenzoxy lysine |
| β-Benzyl aspartate | Leucine |
| β-Benzyl aspartate | Alanine |
| β-Benzyl aspartate | Valine |
| Aspartic acid/β-benzyl aspartate | Leucine |
| β-Benzyl aspartate/β-methyl aspartate | Leucine |
| Glutamic acid | Leucine |
| Glutamic acid | Valine |
| Glutamic acid | γ-Ethyl glutamate |
| Glutamic acid | Lysine |
| ε-Carbobenzoxy lysine | Leucine |
| ε-Carbobenzoxy lysine | Phenylalanine |
| ε-Carbobenzoxy lysine | Alanine |
| Lysine | Leucine |
| Lysine | Phenylalanine |
| Lysine | Alanine |
| D.L-Methionine | Leucine |
| $N^5$-(2-Hydroxyethyl) glutamine | Valine |
| $N^5$-(2-Hydroxyethyl) glutamine | Leucine |
| $N^5$-(2-Hydroxyethyl) glutamine | Glutamic acid |
| $N^5$-(2-Hydroxypropyl) glutamine | Glutamic acid |

Studies in our laboratory have shown that the random copolymerization of the *N*-carboxyanhydrides of two amino acids using a strong base as the initiator in a low dielectric media results in high-molecular-weight copolymers at low conversion.[7] This behavior indicates that these polymerizations are similar to vinyl-type polymerizations and permits the application of the monomer reactivity ratio theory to these polymerizations. This is important as it

## Table 5
## BLOCK COPOLYMERS: A − B OR A − B − A TYPES

| A component | B component | Type |
|---|---|---|
| γ-Benzyl glutamate | Butadiene | A − B − A |
| γ-Methyl glutamate | Butadiene | A − B − A |
| γ-Ethyl glutamate | Butadiene | |
| ε-Carbobenzoxy lysine | Butadiene | A − B − A |
| γ-Benzyl glutamate | Dimethylsiloxane | A − B − A |
| γ-Benzyl glutamate | Leucine | A − B − A |
| γ-Benzyl glutamate | Valine | A − B − A |
| γ-Benzyl glutamate | Tetramethylene oxide | A − B − A |
| Leucine | γ-Benzyl glutamate | A − B − A |
| Valine | γ-Benzyl glutamate | A − B − A |
| γ-Ethyl glutamate | Tetramethylene oxide | A − B − A |
| γ-Methyl glutamate | Tetramethylene oxide | A − B − A |
| γ-Benzyl glutamate | Acrylonitrile/butadiene | A − B − A |
| ε-Carbobenzoxy lysine | Butadiene | A − B |
| ε-Carbobenzoxy lysine | Styrene | A − B |
| Lysine | Styrene | A − B |
| Lysine | Butadiene | A − B |
| γ-Benzyl glutamate | Butadiene | A − B |
| γ-Benzyl glutamate | Styrene | A − B |
| γ-Benzyl glutamate | Carbohydrate fraction of ovamucoid glycoprotein | A − B |
| $N^5$-(3-Hydroxypropyl) glutamine | Butadiene | A − B |
| $N^5$-(2-Hydroxyethyl) glutamine | Leucine | A − B − A |
| $N^5$-(2-Hydroxyethyl) glutamine | Butadiene | A − B − A |
| $N^5$-Hydroxypropyl glutamine | γ-Benzyl glutamate | A − B − A |
| $N^5$-(2-Hydroxyethyl) glutamine | γ-Benzyl glutamate | |

indicates that the copolymer chains which form during the polymerization may have different compositions depending on the reactivities of the respective *N*-carboxy-anhydrides.[8,9] The usual synthetic procedure is to copolymerize various amino acid *N*-carboxyanhydrides to near 100% conversion, isolate the copolymer, and then determine the average amino acid composition of the copolymer. Little attention is generally given to the fact that a large number of chains are formed in which the composition of each individual chain is dependent upon the respective reactivity ratios of the NCA monomers and the monomer mole ratio present in the copolymerization at a given conversion (i.e., reaction time) that the respective copolymer chains are formed. This compositional heterogeneity may be important when determining structure/property relationships. Knowledge of the respective reactivity ratios and use of the copolymer composition equation permits the average and incremental copolymer composition and the monomer feed ratio at any conversion to be determined. Variation in the solvent used for copolymerization may lead to copolymers with variable interchain compositions, and this variation in interchain compositional heterogeneity may be reflected in the solid-state conformations of the respective copolymers. This may also be important in determining structure/property relationships. Polyamino acids are known to adopt certain conformations — α-helix, β-sheet, and random coil — and the ability to adopt certain conformations is dependent on the type of amino acid used in the preparation of the polymer under investigation. We have shown that *N*-carboxyanhydride copolymerizations may exhibit interchain compositional heterogeneity which can contribute to conformational heterogeneity, i.e., chains with varying compositions which exhibit varying amounts of α-helix, β-sheet, and random structures.[9] Obviously, care should be taken in the interpretation of structure/property results where the random copolymer system under investigation may exhibit interchain compositional heterogeneity with conformational heterogeneity. These variations in primary structure (composition) and secondary structure (conformation) may lead to variations in tertiary structure (solid state structure) which may be important in the overall determination of the applicability of a given α-amino acid copolymer as a biomedical material.

## B. Polymer Properties

Polymer properties considered to be important in the use of any polymer for a biomedical application include an understanding of the surface properties, morphology, mechanical properties, and permeability and diffusion. These properties have been investigated for a wide variety of polyamino acids. In most cases, the specific method of characterization was dependent on the general application considered for a given material.

The secondary structure or conformation of the polyamino acid may play a role in determining its specific properties. This was pointed out in 1969 by Baier and Zisman[10] who reported on the influence of polymer conformation on the surface properties of poly-γ-methyl glutamate and poly-γ-benzyl glutamate. Casting these polymers from different solvents and carrying out surface tension measurements and multiple attenuated total reflectance infrared spectroscopic experiments, these investigators determined that the helical and random conformations both showed differences in wetting by hydrogen-bonding and nonhydrogen-bonding liquids, producing a range of critical surface tensions from about 40 to 50 dyn/cm. The β-sheet structure did not exhibit this preferential wetting by hydrogen-bonding liquids and had a single critical surface tension of about 40 dyn/cm or less. Their major conclusion was that the backbone conformation of a poly-α-amino acid, protein, or other biopolymer could strongly influence the surface interactions of that polymer.

Parks et al.,[11,12] using a series of glutamate:leucine and γ-benzyl glutamate:phenylalanine copolymers synthesized in our laboratory, found that the contact angles for all liquids decreased with increasing γ-benzyl glutamate concentration. Correlative blood coagulation studies were carried out and it was determined that the measured partial thromboplastin times were greatest with those materials where the polar component of the critical surface tension was the greatest. Platelet adhesion and deactivation of Coagulation Factor VII also appeared to correlate best with the polar component of the surface tension. In an in vivo experiment using canines, Van Kampen et al.[13-15] showed that no simple relationship existed between the surface properties and the degree of thrombosis resistance. Helmus et al.,[16,17] using a series of glutamic acid:leucine copolymers, showed that these materials implanted in the femoral and carotid arteries of canines gave varying degrees of thrombosis. These investigators were able to show that the amount of thrombus formed on the surface of these copolymer films could be related to the composition of the film and the degree of ionization of the glutamic acid carboxyl groups. When the initial surface concentration of the unionized glutamic acid was greater than 10%, the surface was completely covered with a thrombus which was limited to 600 μm in thickness. For surface concentrations of unionized glutamic acid less than 10%, the amount of thrombus was a linear function of the degree of ionization. When 10% of the total surface sites consisted of ionized glutamic acid residues, there was no thrombus and only formed elements adhered to the surface. These investigators also carried out streaming potential measurements on the glutamic acid:leucine films. Table 6 contains data on the critical surface tension with polar and dispersive components of selected polyamino acids.

## C. Morphology

Major emphasis over the past decade has been on the use of polyamino acids in solid forms as biomedical materials. With this effort has come an increasing interest in the bulk material properties of these various polymers. In particular, efforts directed to determine the bulk and surface morphologies of the polyamino acids has increased.

## Table 6
## CRITICAL SURFACE TENSIONS OF SELECTED POLY-α-AMINO ACIDS

| Polymer | Casting solvent | $\gamma_s$(dyn/cm) | $\gamma_s^p$(dyn/cm) | $\gamma_s^d$(dyn/cm) | Polarity(%) |
|---|---|---|---|---|---|
| γ-Benzyl glutamate | Benzene | 38 | 5 | 34 | 12 |
| γ-Benzyl glutamate/leucine 4:1 | Benzene | 37 | 6 | 32 | 16 |
| γ-Benzyl glutamate/phenylalanine 4:1 | Dioxane | 35 | 6 | 29 | 17 |
| γ-Benzyl glutamate/leucine 1:1 | Benzene | 32 | 6 | 26 | 19 |
| γ-Benzyl glutamate/leucine 1:4 | Benzene | 26 | 2 | 24 | 8 |
| γ-Benzyl glutamate/phenylalanine 1:4 | Dioxane | 28 | 2 | 26 | 7 |
| Glutamic acid/leucine 1:1 | Ethanol | 40 | 15 | 25 | 38 |
| Glutamic acid/leucine 3:2 | Chloroform/10% pyridine | 30 | — | — | 17 |
| Glutamic acid/leucine 7:3 | Chloroform/10% pyridine | 31 | — | — | 17 |
| Glutamic acid/leucine 1:4 | Chloroform/10% pyridine | 31 | — | — | 14 |
| Glutamic acid/leucine 1:9 | Chloroform/10% pyridine | 30 | — | — | 17 |

Following the early work by Keith and co-workers on polyamino acid single crystals, Geil and co-workers have investigated a wide variety of amino acid random and block copolymers.[18-21] They have shown that random copolymers of two or more amino acids are frequently ordered in the solid state. X-ray patterns from copolymers with varying relative amounts of γ-benzyl glutamate and leucine indicate that substantial portions of the macromolecules have an α-helical conformation and have reasonable lateral regularity, the degree of lateral regularity decreasing with increasing leucine content. They have shown that random γ-benzyl glutamate:leucine copolymers cast from solution give fibrils in which the macromolecules are parallel to the fibril axes. Mohadger and Wilkes[22] carried out an extensive study on the effect of casting solvent on the material properties of poly-γ-methyl-D-glutamate. They demonstrated significant differences in the material properties of poly-γ-methyl-D-glutamate when cast from different solvents. Not only was the degree of crystallinity and the molecular conformation affected, but the nature of the bulk material was also altered. These morphological differences influence the material properties in both the wet and dry states. Correlations were made between secondary structure (conformation), tertiary structure (morphology), and the resultant material properties. Experiments included dynamic mechanical measurements, stress-strain measurements, wide-angle X-ray scattering, small-angle light scattering, and infrared spectroscopy.

The capability of block copolymers to give unique material properties has led investigators to prepare and study block copolyamino acids of the A − B and A − B − A types.[23-42] The unique feature of this class of materials lies in the potential for controlling structure on the molecular level — chain conformation and organization, as well as at the level of phase separation. Hayashi et al.[24-27] have prepared and examined the solid state properties of triblock copolypeptides of γ-benzyl glutamate and leucine or valine.[24-27] Wide-angle X-ray diffraction showed a phase-separated morphology which was supported by the results of dynamic mechanical measurements of the viscoelastic behavior of these block copolypeptides. As expected, the choice of solvent for film casting had a significant effect on the solid state structure of the cast films. Films cast from preferential solvents for one phase gave evidence of clear phase separation. The solid state morphology also depended on the position of the soluble block and the properties resembled most closely those of the continuous phase. It was also noted for these block copolymers that phase separation was not as distinct in films cast from a nonpreferential solvent, i.e., a solvent which does not discriminate between the two phases, and in this case, the properties were closer to those of random copolymers.

Barenberg et al.[23,28] have synthesized and characterized two triblock A − B − A copolymers of γ-benzyl glutamate (A) and butadiene/acrylonitrile (B). Solvent studies of the block copolymers revealed that dioxane was a preferential solvent for the γ-benzyl glutamate segment and chloroform was a nonpreferential solvent in that it solvated both the γ-benzyl glutamate and butadiene/acrylonitrile segments. X-ray diffraction and Fourier transform infrared spectroscopy of films cast from dioxane and chloroform showed the γ-benzyl glutamate segments to be predominantly α-helical and disordered α-helical, respectively. Detailed examination of the solvent effect on solid state properties and morphology of the block copolymers was carried out using electron microscopy, tensile testing, and dynamic mechanical and dielectric spectroscopy. Electron microscopy of osmium tetroxide-stained films cast from dioxane revealed lamellar domain formation indicative of phase separation. The midblock butadiene layers were approximately 150 Å while the alternating γ-benzyl glutamate layers were 300 and 500 Å thick for the respective copolymers. Films cast from chloroform exhibited a nearly homogeneous morphology indicative of considerable phase mixing. Tensile testing of films cast from dioxane and chloroform added additional support to the block nature of the copolymers and the morphological dependence of their properties. Dynamic mechanical spectroscopy indicated that the side-chain transition of the γ-benzyl glutamate appeared as a single peak when the copolymers were cast from dioxane (preferential solvent), and a double peak appeared when copolymers were cast from chloroform (nonpreferential solvent).

Block copolyamino acids of the A – B type have been synthesized by Gallot and co-workers. These investigators have carried out extensive solid state characterization of these polymers. The large majority of A – B block copolymers listed in Table 5 have been prepared and characterized by Gallot et al.[29-31]

Nakajima and co-workers[32-40] at Kyoto University have directed a major effort toward the synthesis and characterization of block copolyamino acids of the A – B – A type. Block copolymers investigated by this group have usually contained an ester derivative of glutamic acid (γ-methyl, γ-ethyl, or γ-benzyl). In addition to the usual types of characterization of the block copolymers, Nakajima and co-workers have investigated domain formation and domain structures of their block copolymers using thermodynamic theory and electron microscopic observations. They have extended their studies to include, in some cases, an examination of the hydraulic permeability of their materials and have carried out in vivo tests related to thrombus formation and tissue interactions.[34] On the basis of their antithrombogenicity studies, these investigators suggest that the antithrombogenicity of the materials investigated was due, in part, to their microheterophase structure and domain dimensions. Their materials exhibited lamellar or cylindrical domains ranging from 300 to 450 Å. Their animal studies revealed antithrombogenic behavior for their γ-ethyl glutamate:butadiene:γ-ethyl glutamate A – B – A block copolymer. Further studies with these block copolymers using the adsorption of proteins from human blood as estimated by immunoelectron microscopy showed that the one block copolymer which showed antithrombogenic behavior in vivo exhibited only the adsorption of albumin onto its surface. The other two block copolymers containing γ-ethyl glutamate and butadiene did not show an antithrombogenic behavior in vivo and exhibited a variable adsorption of proteins from human blood.

## D. Mechanical Properties

Relatively little effort has been directed toward examining the stress-strain properties of polyamino acids. Anderson et al.[43,44] examined the tensile properties of γ-benzylglutamate/leucine copolymer films and found that as the leucine content increased, the tensile strength, extension at fracture, and modulus increased. The respective copolymer films also were treated in pseudoextracellular fluid for 48 hr at 100° and tested. In general, this treatment led to an increase in tensile strength and modulus and a decrease in elongation at break. Detailed studies were carried out on a 50 mol % γ-benzylglutamate/50 mol % leucine copolymer. While treatment with pseudoextracellular fluid increased the yield stress, acid or enzyme treatments were comparable to the wet control in terms of the yield stress and modulus. These results suggested that the copolymer film was highly impervious to such treatments and the environmental effect may be plasticization by water rather than chemical modification of the polymer.

Mohadger and Wilkes[22] have examined the effect of the casting solvent on the material properties of poly-γ-methyl-D-glutamate. The use of different casting solvents resulted in different conformations or secondary structures of the chains in the poly-γ-methyl-D-glutamate. Table 7 shows the Young's modulus in the wet and dry films of poly-γ-methyl-D-glutamate cast from various solvents. The dependence of Young's modulus on the conformation is obvious.

As tensile properties of polymers are usually related to molecular relaxations at or near the temperature of interest, a knowledge of the relaxation behavior of the polymer is important in possible design considerations. For this reason, dynamic mechanical testing is of interest.

Numerous investigators have studied the molecular relaxations in polyamino acids by examining the dynamic mechanical relaxation behavior. Anderson et al.[43] have related the bulk tensile properties of copolymers of γ-benzyl glutamate and leucine to their respective

**Table 7**
**PROPERTIES OF POLY-γ-METHYL-D-GLUTAMATE FILMS**

| Casting solvent | Predominant conformation | Young's modulus, E (dyn/cm² × 10⁻⁹) | |
| --- | --- | --- | --- |
| | | Dry film | Wet film |
| Chloroform | α-Helix | 11.4 | 3.6 |
| Trifluoroacetic acid | Random coil | 8.2 | 2.6 |
| Formic acid | β-Sheet | 18.7 | 2.1 |

molecular relaxation behavior. Mohadger and Wilkes[22] have correlated the relaxation behavior of poly-γ-methyl-D-glutamate with the chain conformation and the morphology of the specimen.

It is of interest that polyamino acids with large side-chain substituents such as γ-benzyl glutamate or ε-carbobenzoxy lysine exhibit a major molecular relaxation at or slightly below physiological temperatures.[45] This relaxation has been attributed to side-chain motion. Incorporation of monomeric units with large side-chain substituents such as the γ-benzyl glutamate, γ-methyl glutamate, or ε-carbobenzoxy lysine in copolymers result in films which are pliable rather than stiff or brittle.

Dynamic mechanical techniques have been used to investigate block copolyamino acids. Hayashi et al.[27] have examined the dynamic mechanical behavior of block copolymers of γ-benzyl glutamate and leucine and have found that the viscoelastic behavior supports the suggested phase-separated morphology. Representation of the dynamic elastic modulus in terms of the equivalent mechanical model suggested that phase separation of glutamate-leucine-glutamate A − B − A-type block polymers occurs with glutamate as the matrix phase. It should be noted that as the morphology of block polyamino acids is dependent on casting solvent, the observed viscoelastic behavior of these block polymers is also dependent on casting solvent.

An interesting study relating chain conformation and the physical and mechanical properties of a cross-linked poly(ω-hydroxyalkyl) glutamine has been carried out by Sugie and co-workers.[46-49] The mechanical response of the swollen polymer varied markedly from elastic behavior in water to plastic deformation in methanol. The change in deformation mechanism was attributed to the difference in chain conformation. In water, the chain is in the highly flexible random coil conformation. As a result, the behavior is typically that of an elastomer with the predicted relationships between the swelling ratio and cross-link density, the low modulus, and the reversible, time-independent mechanical response. In methanol, the chain is in the α-helix conformation and the mechanical behavior was that of a plastic with time-dependent, irreversible deformation.

## E. Permeability and Diffusion

In considering poly-α-amino acids for biomedical materials, an understanding of their transport properties is important. Oohachi et al.[50] have investigated the permeation and sorption of carbon dioxide with the poly-γ-benzyl glutamate films which were prepared under different casting temperatures and different casting solvents. The different film casting methods produced films which had variable intra- and intermolecular stacking of the benzene rings and the side-chain. These investigators correlated the degree of side-chain benzene ring stacking and the permeation and sorption behavior of carbon dioxide in the various films. A decrease in stacking caused an abrupt increase in permeability and an increase in the amount of sorption. Kawai et al.[51] investigated the sorption and permeability of carbon dioxide and oxygen in films of polylysine which had varying degrees of side-chain carbobenzoxylation. They found that the ratio of the solubility of carbon dioxide to oxygen decreased remarkably with decreasing content of the carbobenzoxy side-chain group.

The permeability properties of leucine/methionine copolymers have been investigated by Martin et al.[52] and Klein et al.[53] These investigators used peroxide oxidation of the side-chain methyl-thioethyl groups to methyl-sulfinoethyl groups to enhance the hydrophilic properties of the membranes and the water vapor transmission rates. They found that increasing the leucine content of the copolymer membranes produced increases in both the carbon dioxide permeability and the oxygen permeability. Increasing the methionine content increased the carbon dioxide-oxygen separation factor. Of special interest is the relationship between morphology, molecular relaxations, and permeation behavior. Many of the random and block copolymers contain large side-chain substituents which give major molecular relaxations at or close to physiological temperatures; this behavior alters greatly the observed diffusion or permeability behavior.

## III. BLOOD COMPATIBILITY

### A. Coagulation

Numerous studies have been carried out on the interaction of polyamino acids with blood and its components; however, these studies have, for the most part, involved water-soluble polypeptides and have been focused on the ability of synthetic analogs to reproduce structural features of proteins and/or the ability of synthetic analogs to reproduce or induce the biological function of proteins. This effort was concentrated in the 1950s and 1960s and has been extensively reviewed. Only in the 1970s was attention given to the interaction of insoluble polyamino acids and their potential as blood-contacting materials. For the sake of simplicity, this review will address separately the interaction of insoluble polypeptides in coagulation, with platelets, and in thrombosis.

In 1975, Komai and Nose[54] published information on the whole blood compatibility of glutamate ester/leucine copolymers. The relative compatibility determined by an in vitro modified kinetic method increased in the order poly-γ-methyl-D-glutamate, copoly(γ-benzylglutamate:γ-methyl-D-glutamate) 1:4, copoly(γ-methylglutamate:leucine) 1:1, copoly(γ-benzylglutamate:leucine) 1:4, copoly(γ-benzylglutamate:leucine) 1:1, poly-γ-benzylglutamate.

In the mid 1970s, the group at Case Western Reserve University made a concentrated effort to investigate blood-material interactions using insoluble polyamino acids. Parks et al.[11,12] studied the role of bulk and surface properties of selected amino acid copolymers on blood coagulation and determined that the partial thromboplastin clotting times of the respective copolymers was dependent on the choice of casting solvent and the subsequent morphology of the cast films. When cast from benzene or dichloroethane, copoly(γ-benzylglutamate:leucine) 1:4 gave longer partial thromboplastin times than siliconized glass. Copoly(γ-benzylglutamate:leucine) 1:1 exhibited the shortest partial thromboplastin clotting times when cast from benzene, dichloroethane, or dioxane. Factor VII assays showed that the effect of composition was significant and the copoly(γ-benzylglutamate:leucine) 1:1 showed the greatest clotting rates and relative bioactivity.

In vitro clotting experiments have been carried out on various amino acid copolymers.[55] Table 8 gives the data for the respective clotting times of whole blood and plasma as measured by the partial thromboplastin times. For the hydrophobic copolymers, there appears to be little effect on the intrinsic clotting system.

Walton and co-workers[56] have examined interaction between blood proteins and coagulation proteins with numerous polyanionic, polycationic, hydrophilic, and hydrophobic polyamino acids and their derivatives. Table 9 shows a summary of these interactions. The conclusions reached by this group are that although residual charge plays an important role, particularly in solution, the composition of the polyamino acid film appears to play a relatively

**Table 8**

**EFFECT OF AMINO ACID COPOLYMERS ON BLOOD CLOTTING
IN VITRO**

| Copolymer | Composition | PTT times for plasma (sec) | Whole blood clotting times (min) |
|---|---|---|---|
| Glutamic acid:leucine | 1:4 | 87 | — |
| Glutamic acid:leucine | 1:1 | 81 | 14 |
| γ-Benzyl glutamate:leucine | 1:1 | 153 | 29 |
| γ-Benzyl glutamate:phenylalanine | 1:1 | 285 | 23 |
| γ-Benzyl glutamate:valine | 1:1 | 255 | 33 |
| γ-Benzyl glutamate:alanine | 1:1 | 292 | 34 |
| ε-Carbobenzoxy lysine:leucine | 1:1 | 243 | 45 |
| Glass | | 80 | — |

**Table 9**

**BLOOD PROTEIN INTERACTION WITH AMINO ACID
COPOLYMERS CONTAINING GLUTAMATE OR
LYSINE MONOMER UNITS**

| Blood protein | Glutamate | Lysine | Comment |
|---|---|---|---|
| Factor XII Hageman factor | X | | Copolymer causes enzyme activation |
| Factor X Stuart factor | X | | Conformational change and/ or binding |
| Factor IX Christmas factor | X | | |
| Factor V Proaccelerin | X | X | Binding with both polymers |
| Factor II Prothrombin | | X | Conformation effect on protein and activation |
| Thrombin | | X | |
| Fibrinogen | X | X | |
| Albumin | | X | |
| γ-Globulin | | X | |

minor part in their interactions with blood components, and features such as flexibility, hydration, surface mobility, and substrate morphology become important.

Protasi et al.[57] have investigated the in vitro effects on blood coagulation of poly-($N^5$-(2-hydroxyethyl)glutamine) for possible use as a synthetic plasma substitute. This nonionic, water-soluble polyamino acid was similar to Dextran in its behavior and decreased the coagulation time in the intrinsic and extrinsic coagulation systems and in the thrombin-fibrinogen system. These investigators suggest that the random coil structure of the hydroxyalkyl glutamine polymer probably facilitates contact with the fibrinogen and fibrin monomers and subsequently induces the mechanism of steric exclusion which is responsible for the procoagulating and precipitating events.

## B. Platelet Interaction

Little attention has been given to the interaction of blood platelets with polyamino acid films. In 1979, Solomon and co-workers[58,59] published their extensive work on the interaction of washed human platelets with films of amino acid copolymers. They concluded from kinetic and equilibrium studies of blood platelet binding to amino acid copolymer films that attachment and serotonin release were not dependent upon the composition of the amino

**Table 10**
## PLATELET ADHESION AND RELEASE ON AMINO ACID COPOLYMER FILMS

| Copolymer | Copolymer composition | Adhesion (%) | Release (%) |
|---|---|---|---|
| Glass | | 30 | 12.6 |
| γ-Benzyl glutamate | | 19 | 18.4 |
| γ-Benzyl glutamate:leucine | 4:1 | 27 | 18.1 |
| γ-Benzyl glutamate:leucine | 1:1 | 43 | 20.8 |
| γ-Benzyl glutamate:leucine Albumin precoated | 1:1 | 14 | 3.9 |
| γ-Benzyl glutamate:leucine Fibrinogen precoated | 1:1 | 10 | 4.6 |
| γ-Benzyl glutamate:leucine γ-Globulin precoated | 1:1 | 16 | 5.6 |
| γ-Benzyl glutamate:leucine | 1:4 | 33 | 18.1 |
| Glutamic acid:leucine | 1:4 | 41 | 25.8 |
| Lysine:leucine | 1:99 | 41 | 18.3 |
| Lysine:leucine | 1:9 | 32 | 16.6 |
| Lysine:phenylalanine | 1:4 | 38 | 13.2 |

acid copolymer. Table 10 shows the adherence and serotonin release in their kinetic system for a range of acidic, hydrophobic, and basic amino acid copolymers. Also included in Table 10 is information on platelet interactions after the amino acid copolymer has been precoated with albumin, fibrinogen, and γ-globulin.

## C. Thrombosis

Several groups have carried out in vivo studies utilizing polyamino acids. These studies have been directed towards assessing the effect of different chemical moieties upon arterial thrombosis, and the three groups involved in these studies utilized the canine model.

Barenberg and co-workers[60-63] examined block copolymers: γ-benzylglutamate/acryloni-trile-butadiene/γ-benzylglutamate and hydroxypropylglutamine/γ-benzylglutamate/hydrox-ypropylglutamine. These block copolymers were studied on a rotating shaft in an extracorporeal circulation from dogs. From his studies, Barenberg constructed a semiempirical epitaxial model which correlated and interrelated the surface free energy, ultrastructural morphology, surface charge, surface chemistry, and surface molecular motions of the triblock copolymer to thrombogenesis. In particular, Barenberg and co-workers have utilized these model block copolymers and results from studies in vivo to address the specific questions of the effect of structured water, and macromolecular motions and order of the polymer interface on hemocompatibility.

Van Kampen et al.,[13-15] utilizing a series of glutamate/leucine copolymers, showed that changes in implant surface chemistry elicited a range of responses in the carotid arteries of dogs that varied from intense thrombosis and rapid vessel occlusion to minimal thrombosis and endothelialization. They found no simple relationship between the gross surface property, such as hydrophobicity, and the degree of thrombosis resistance. Some hydrophobic and hydrophilic materials were found to have good thrombosis resistance while others were found to have poor thrombosis resistance. Leukocytes were shown to play an important role in both the initial thrombosis and endothelialization on the glutamate/leucine copolymers. The major difference between the copolymers that progressed to rapid vessel occlusion and materials that remain patent was the degree of direct leukocyte adherence and spreading on the surface of the amino acid copolymer prior to extensive platelet aggregation. Amino acid copolymers exhibiting a lack of direct leukocyte adherence to the surface were associated with intense thrombosis and rapid vessel occlusion. Following thrombus thickness and composition as a function of implantation time, these investigators constructed a hypothetical model representing the sequence of events and alternative pathways occurring at the blood-material interface. The model specifically included the involvement of leukocytes in arterial thrombosis.

Nakajima and co-workers[32-40,64] have concentrated on investigating the role of micro-heterophase structures on the blood-materials interaction. Nakajima has proposed that the interaction of platelets with artificial materials is dependent upon the microheterophase structure. In addition, he has suggested that when a material having a microheterophase structure is placed into flowing blood, adsorption of plasma protein is controlled by the microheterophase structure of the material surface. He has further suggested that micro-heterophase structure reflecting the surface structure of the material may be realized with respect to the adsorbed protein layer. Nakajima and co-workers have carried out extensive studies using block copolyamino acids and the results from studies on these materials suggest that these hypotheses may have validity.

## IV. TISSUE RESPONSE

### A. Inflammatory Reaction

To specifically examine the inflammatory response, i.e., in vivo biocompatibility, of a biodegradable hydrogel, poly(2-hydroxyethyl glutamine), Marchant and co-workers[65] have utilized a cage implant system which permitted the in vivo evaluation of the components of the inflammatory reaction and the activation of inflammatory cells. A comparison of the cellular response for the PHEG system and the control system did not show statistically significant differences during the first 7 days following implantation. The acute inflammatory response predominated by polymorphonuclear leukocytes was followed by a mild chronic inflammatory response where macrophages and lymphocytes were predominant. During this chronic phase, 8 to 14 days following implantation, macrophages were present in significantly larger numbers for the PHEG system when compared to the control values. Enzyme analysis of the exudates revealed statistically significant differences between the PHEG system and the control system at time intervals where no differences were noted in cell density or population. Stress-strain measurements on implanted PHEG samples showed that significant in vivo degradation had occurred during the acute inflammatory phase of the reaction, i.e., the first 7 days following implantation.

The cage system permits the evaluation of biocompatibility over a relatively short period of time following implantation, i.e., 21 to 28 days. As the anticipated use of polyamino acids as biomaterials is for extended periods of time, long-term implantations have been carried out with these polymers and copolymers. Emphasis in these studies has been placed on evaluating the formation of the fibrous capsule which is the end result of the inflammatory reaction.

### B. Fibrous Capsule Formation

Anderson et al.[44] have studied the tissue compatibility of esterified glutamic acid copolymers implanted for periods ranging up to 200 days. The appearance of the fibrous capsule at 200 days for polymers containing benzyl or methyl esters was similar. Grossly, a thin, translucent, fibrous capsule surrounded the implant with the implant being freely movable within the capsule. The capsule consisted of collagen fibers and occasional fibroblasts, and some macrophages were noted at the interface. Hydrophobic amino acid homopolymers and copolymers showed essentially the same response upon implantation. By 30 days following implantation, the capsule is well developed and vascularity of the capsule is at a minimum. In these studies, little difference was seen in the biological response to the variable composition of the polymers.

Chen et al.[34] have examined the tissue compatibility of γ-ethylglutamate/butadiene/γ-ethylglutamate block copolymers. These materials were examined by coating the block copolymers on Dacron fabric which was subsequently implanted in rabbits for 4 weeks. These materials were considered to exhibit good tissue compatibility as the tissue reaction was similar to that seen for the Dacron fabric control.

Noishiki et al.[64] have examined the tissue reaction to A − B − A block copolyamino acids composed of polyamino acids as the A component and polybutadiene or polytetramethylene oxide as the B component. These materials exhibited a minimal tissue reaction and were considered to show good tissue compatibility.

## C. Cell Interactions

The kinetics of attachment of NIL B and SV-NIL cells to glass surfaces coated with random copolyamino acids, glass, and siliconized glass have been studied by Soderquist et al.[66-68] They found that in the absence of serum proteins, neither the rate or extent of the attachment of cells were affected by the nature of the surface. In the presence of bovine serum albumin, the total uptake and rate of attachment of both NIL B and SV-NIL cells to the neutral, hydrophobic, and negatively charged copolyamino acids were decreased compared with attachment to the same surfaces in the absence of protein. In contrast, the attachment of NIL B and SV-NIL cells to the positively charged lysyl copolymers was not decreased in the presence of protein. The results from this study support the concept that while cells bind to an adsorbed layer of protein rather than directly to the surface, the underlying surface can modify the attachment process by its effect on the protein adsorbed.

Important in determining the biocompatibility of candidate biomedical polymers is the study of their cytotoxicity in vitro. Sudilovsky and co-workers[69] have examined the in vitro cytotoxicity and in vivo acute responses to some amino acid copolymers and their esters. These investigators used mouse L-929 fibroblasts in an in vitro agar overlay cell culture test. A wide range of hydrophilic and hydrophobic copolymers were examined and compared with in vivo subcutaneous testing in rats. While a wide variation in the type and number of inflammatory cells was observed around the implants, only necrosis was identified in vivo whenever a positive cytotoxic response was present with the L-929 cells in vitro.

Polyamino acids with basic groups in their side chains, i.e., polylysine, polyornithine, and polyhistidine, are frequently used to treat surfaces to enhance cell attachment and subsequent cell growth.[70] This is a commonly accepted technique in cell culture studies.

# V. BIODEGRADATION

Hydrophilic polyamino acids have been known for some time to be biodegradable. This section will only deal with the biodegradation of polyamino acids which are in the film form. In 1974, Anderson et al.[44] studied a series of copolymers in which the γ-carboxyl group of glutamic acid was not esterified. In general, the higher the percentage of free glutamic acid in the copolymer the faster the implant degraded. The amount of polmeric material remaining in vivo after 14 or 30 days implantation was variable and this variation was attributed to differing thicknesses of the film implants. The inflammatory response of these materials compared to their parent γ-benzylglutamate copolymers was increased and the presence of lymphocytes and macrophages, as well as foreign body giant cells, were noted for longer implantation time periods. A film of polyglutamic acid was implanted and was noted to have completely disappeared by 14 days.

Dickinson et al.[71,72] studied the in vitro and in vivo degradation of the hydrophilic, nonionic polyamino acid, poly(2-hydroxyethyl glutamine). The in vivo biodegradation of specimens implanted in the peritoneal cavity of rats was followed by the variation in the swelling ratio of the polymer. Degradation of the polymer was observed only during the first 2 weeks of implantation and was attributed to hydrolysis by proteolytic enzymes released during the acute and chronic stages of the normal inflammatory response. Following on the in vivo studies, these investigators carried out in vitro model studies using several enzymes. Trypsin and collagenase had no effect on the cross-linked PHEG but pronase and papain degraded the hydrogel. The initial effect of papain was to decrease the effective cross-link density without producing soluble material. This effect was similar to that observed in vivo. As previously described, this polymer also has been tested in the in vivo cage system for its ability to modulate the acute inflammatory response.[65] Hayashi and co-workers[73,74] have examined the biodegradation of polyamino acids in vitro using enzymes.

Marck et al.[75] have examined the biodegradability and tissue reaction of a series of random copolyamino acids with varying percentages of hydrophilic aspartic acid and hydrophobic monomers such as leucine, β-methylaspartate, and β-benzylaspartate. These studies were carried out in rats which had been implanted with films of these materials subcutaneously. These investigators noted that three groups of materials with different ranges of hydrophilicity could be distinguished: hydrophobic materials which showed no degradation after 12 weeks of implantation; more hydrophilic materials which revealed a gradual reduction in size of samples but were still present after 12 weeks implantation; hydrophilic copolymers which disappeared in 24 hr. They noted that the tissue reactions for the materials which did not undergo biodegradation were similar to that observed for silicone rubber, but the capsules were adherent to the surfaces of the polyamino acids in contrast to silicone rubber. With the materials which underwent slow biodegradation, the tissue reaction was marked by the occurrence of lymphocytes and plasma cells and increased numbers of macrophages, foreign body giant cells, and eosinophilic leukocytes. The materials which disappeared within 24 hr showed only a temporary chemical inflammatory response.

## VI. ANTIGENICITY OF POLY-α-AMINO ACIDS

Of major concern in considering polyamino acids as biomedical materials is the fact that these are polypeptide or protein-like macromolecules and as such may elicit an immune response when injected or implanted into mammals. The use of polyamino acids as antigens for immunogenicity studies is well established. Obviously, it is not the purpose of this review to summarize the extensive work that has gone on in this area but rather to present a perspective on the immunogenic or nonimmunogenic behavior of polyamino acids.

Although some polyamino acid homopolymers and copolymers have been shown to be antigenic, terpolymers of amino acids containing three or more different types of amino acids have been found to produce the most antigenic responses. It should be noted that polymers containing three or more different types of amino acids are not generally considered as being candidates for biomedical materials. No such materials have been described in this review. In referring to the antigenicity of a polymer, we are usually referring to the ability of the polymer to cause the production of a circulating antibody in the host.

Important in determining the host reaction to a given polyamino acid is the species in which the polyamino acid is tested. Maurer[76-80] has shown that copolymers of glutamic acid and lysine can produce antibodies in rabbits while similar intramuscular injections produce no detectable antibody in man.

As previously noted in the structure segment of this review, amino acids may have the L or D configuration. While the vast majority of studies presented in the review have utilized L-amino acids, Maurer as well as others have shown that the use of D-amino acids in synthetic polyamino acid copolymers and terpolymers can markedly alter the antigenic response. In general, the use of a D-amino acid will decrease the antigenicity of a polyamino acid.

## VII. APPLICATIONS

Poly-α-amino acids have been investigated for possible use in a wide variety of biomedical applications. The applications have ranged from biodegradable sutures to matrix materials and carriers for drug delivery systems.

The development of filaments and surgical sutures of partially esterified poly-L-glutamic acid has been explored by Miyamae and co-workers.[81] Sutures of poly-L-glutamic acid, in which the outer surfaces of the filaments were partially esterified with ethanol, were examined in regard to degree of esterification and absorption time. Sutures ranging from ca. 7 to 54% esterification and 2 to 60 days absorption time, respectively, were tested in the intestinal tracts of dogs. The absorption time, durability, and tensile properties of these sutures could be controlled by varying the degree of esterification and predrawing of the fibers prior to partial esterification. Copolymer sutures composed of L-glutamic acid and L-valine or L-alanine were also investigated.

Poly-α-amino acid copolymers have been used as artificial skin substitutes in burn therapy.[82,85] In a pilot study, Spira et al.[82] reported that nylon-velour polypeptide laminates examined as artificial skin grafts to experimental burn wounds were well tolerated in clean wounds and served as a framework in which fibroblastic proliferation occurred with adherence or take. In wounds experimentally contaminated with infectious organisms, the polypeptide laminates allowed escape of the infectious exudate and delayed separation of the laminate from the wound bed. Bacterial counts from burn wounds with grafts of the polypeptide laminate were similar to those obtained with autogenous skin grafts. This pilot study was followed by the presentation of three clinical cases involving the use of the polypeptide laminate as an artificial skin.[83] These cases showed the successful temporary use of polypeptide laminates as wound covering prior to autogenous grafting.

Hall and co-workers[85] followed these studies with an evaluation of the nylon-velour polypeptide laminate containing a topical antibacterial agent, silver sulfadiazine.[85] The polypeptide dressing compared well with autografts in terms of ''take'' rates and appeared to retard fluid, electrolyte, and protein loss. The inclusion of silver sulfadiazine resulted in a delayed onset and decreased incidence of Gram-negative, organism-induced wound sepsis.

Martin et al.[52] have reported on the unusual permeability characteristics of DL-methionine-L-leucine copolymer membranes and suggest their possible use in artificial kidney machines and blood oxygenators. Oxidation of the methionine side chain, $-CH_2-CH_2-S-CH_3$, to the corresponding sulfoxide and sulfone increased the hydrophilic nature and water vapor transmission rates of these films. It was also shown that the carbon dioxide-oxygen separation factor could be varied by controlling the DL-methionine to L-leucine ratio; increasing the percent composition of DL-methionine increased the separation factor.

Klein et al.[53] have reported on the dialysis characteristics of DL-methionine-DL-leucine copolymer films. The transport of sodium chloride, urea, creatinine, uric acid, and bacitracin was found to be dependent on the degree of hydration of the polymers, which in turn was a function of the extent of methionine oxidation. For a given copolymer composition, 38 mol % DL-leucine/62 mol % DL-methionine, variable oxidation of the methionine residues gave membranes in which the gel swelling ranged from 12 to 88%. Transport properties were altered by two mechanisms: variable side-chain polarity and variable polymer hydration.

Poly-α-amino acids have been used as carriers for therapeutic agents. Rowland and co-workers[86] have used polyglutamic acid in an antibody-directed carrier system to examine the cytotoxic effect of the antibody-carrier-drug conjugate on suppression of tumor growth in mice. Shen and Ryser[87] have shown that the conjugation of poly-L-lysine to albumin and horseradish peroxidase enhanced the cellular uptake of these proteins. Following up on these studies, these authors showed the increased transport of methotrexate into cells when methotrexate was conjugated to poly-L-lysine. They were able to demonstrate that the conjugation of methotrexate to polylysine overcame drug resistance in cultured cells.[88,89]

Petersen et al. and Anderson et al. have studied in some detail the binding of contraceptive steroids to polyglutamic acid and polyhydroxylalkyl glutamines. Hydrolytic release of the contraceptive steroids from these polymers has been shown to occur over extended periods of time ranging to months when these systems were tested in vitro and in vivo. These systems make use of the labile bond/spacer group concept in linking the drug as a side-chain moiety to the polymer backbone. Conceptually, the labile bond is hydrolyzed, first releasing the drug, followed by hydrolysis of the polymer backbone and subsequent biodegradation of the polymer at later periods of time.[90-96]

The Petersen group has preferred to couple free hydroxyl groups on the steroids to hydroxyalkyl glutamines by use of a carbonate bond. Anderson et al. have utilized the oximino derivatives of steroids and coupled them to the free carboxyl groups of glutamic acid. Petersen et al. have carried out extensive studies in the in vivo effects of varying the extent of drug loading, particle size, and length of spacer group on release behavior. Some of these systems have been noted to release drug for periods beyond 144 days.

Sidman and co-workers have used biocompatible, biodegradable copolymers of glutamic acid and ethyl glutamate as matrix- and membrane-controlled release systems. They have examined the release behavior of a wide variety of drugs, including norgesterel, progesterone, primaquine, *cis*-platinum, and naltrexone. When fabricated into matrix rods or capsules, these copolymers have been used to release drugs in animals at constant rates for prolonged periods of time. These investigators noted that the physical dimensions and copolymer compositions of the different dosage forms could be readily varied to meet specific delivery rate and duration objectives.[97-101] Dohlman et al.[102] have reported on the use of a hydrocortisone-impregnated polypeptide as a sustained release system for the suppression of corneal xenograft inflammatory reactions in rabbits.

Tani et al.[103] have investigated the unusual behavior of a cortisol-polyglutamic acid system where the steroid release was governed by hydrophilic/hydrophobic control.[103] Although the hydrolysis of the labile oximino bond between the cortisol and the side-chain carboxyl group of polyglutamic acid exhibited first-order kinetics, the overall release rate of drug from these insoluble systems was zero order, i.e., constant for time periods extending beyond 100 days. These investigators developed a model to explain the hydrophilic/hydrophobic control of the release behavior.

Feijen and co-workers[104] have carried out an extensive study on the binding of adriamycin to polyglutamic acid. They have examined the influence of spacer group length. Cell interaction studies with these materials show that they exhibit cytotoxic behavior, which appears to be related to the length and type of the spacer group. Their studies point out the importance of the spacer group and how control of the biologic behavior of these conjugates can be achieved through the use of different spacers and the length of the spacer group.

In summary, the polyamino acids as a class of polymers offers a wide range of chemical, physical, mechanical, and biological properties for potential use as biomedical polymers. As our knowledge of tissue-material or blood-material interactions increases, selected polymers composed of amino acids will probably prove to be useful as biomedical polymers.

## ACKNOWLEDGMENT

This effort was supported in part by National Institutes of Health Grants HL-25239 and HL-27277 and an NIH Research Career Development Award, HL-00779, to J. M. A.

# REFERENCES

1. **Bamford, C. H., Elliott, A., and Hanby, W. E.,** *Synthetic Polypeptides,* Academic Press, New York, 1956.
2. **Katchalski, E. and Sela, M.,** *Advances in Protein Chemistry,* Vol. 13, Academic Press, New York, 1958, 243.
3. **Stahmann, M. A., Ed.,** *Polyamino Acids, Polypeptides and Proteins,* Univ. of Wisconsin Press, Madison, Wisconsin, 1962.
4. **Fasman, G. D., Ed.,** *Poly-α-Amino Acids,* Marcel Dekker, New York, 1967.
5. **Walton, A. G.,** *Polypeptides and Protein Structure,* Elsevier/North-Holland, New York, 1981.
6. **Block, H.,** *Poly(γ-Benzyl-L-Glutamate) and Other Glutamic Acid Containing Polymers,* Gordon & Breach, New York, 1983.
7. **Sederel, W., Deshman, S., Hayashi, T., and Anderson, J. M.,** The random copolymerization of γ-benzyl-L-glutamate and L-valine N-carboxyanhydrides: reactivity ratio and heterogeneity studies, *Biopolymers,* 17, 2835, 1978.
8. **Deshmane, S., Hayashi, T., Sederel, W., and Anderson, J. M.,** The influence of interchain compositional heterogeneity on the conformation in random copolymers of γ-benzyl-L-glutamate and L-valine, *Biopolymers,* 17, 2851, 1978.
9. **Mitra, S. B., Patel, N. K., and Anderson, J. M.,** Interchain heterogeneity in 2-hydroxyethylglutamine-L-valine random copolymers, *Int. J. Biol. Macromol.,* 1, 55, 1979.
10. **Baier, R. E. and Zisman, W. A.,** The influence of polymer conformation on the surface properties of poly(γ-methyl-L-glutamate) and poly(benzylglutamate), *Macromolecules,* 3(1), 70, 1970.
11. **Parks, P. J., Jr., Gibbons, D. F., and Malhotra, O. P.,** The role of bulk and surface properties of selected amino acid copolymers on blood coagulation, *Ann. Biomed. Eng.,* 7, 411, 1979.
12. **Parks, P. J., Jr.,** Coagulation Studies on Selected Copolymers of γ-L-Benzylglutamate, L-Leucine and L-Phenylalanine, Ph.D. thesis, Case Western Reserve University, Cleveland, Ohio, 1977.
13. **Van Kampen, C. L., Gibbons, D. F., and Jones, R. D.,** Effect of implant surface chemistry upon arterial thrombosis, *J. Biomed. Mater. Res.,* 13, 517, 1979.
14. **Van Kampen, C. L.,** Effect of Implant Surface Chemistry upon Arterial Thrombosis and Endothelialization, Ph.D. thesis, Case Western Reserve University, Cleveland, Ohio, 1977.
15. **Van Kampen, C. L., Gibbons, D. F., and Jones, R. D.,** The effect of surface chemistry on arterial thrombosis, *Polym. Prepr.,* 20, 36, 1979.
16. **Helmus, M. N., Gibbons, D. F., and Jones, R. D.,** Surface analysis of a series of copolymers of L-glutamatic acid and L-leucine, *J. Coll. Inter. Sci.,* 89(2), 567, 1982.
17. **Helmus, M. N.,** The Effect of Surface Charge on Arterial Thrombosis, Ph.D. thesis, Case Western Reserve University, Cleveland, Ohio, 1980.
18. **Geil, P. H., Barenberg, S., and Wong, W. M.,** Structure, morphology, and mechanical properties of bioplastics, in *Proceedings of the First Cleveland Symposium on Macromolecules,* Walton, A. G., Ed., Elsevier, Amsterdam, 1977, 105.
19. **Tajima, Y., Anderson, J. M., and Geil, P. H.,** Single crystals of a random copolypeptide composed of γ-benzyl-L-glutamate and L-phenylalanine, *Int. J. Biol. Macromol.,* 1, 89, 1979.
20. **Tajima, Y., Anderson, J. M., and Geil, P. H.,** Morphology and structure of γ-benzyl-L-glutamate copolymerized with L-phenylalanine, L-valine and L-alanine, *Int. J. Biol. Macromol.,* 2, 186, 1980.
21. **Wong, W.-M. T.,** The Macromolecular Organization and Motion of Biomembranes, Ph.D. thesis, Case Western Reserve University, Cleveland, Ohio, 1974.
22. **Mohadger, Y. and Wilkes, G. L.,** The effect of casting solvent on the material properties of poly(γ-methyl-D-glutamate), *J. Polym. Sci. Polym. Phys. Ed.,* 14, 963, 1976.
23. **Barenberg, S.,** Structure and Properties of Selected Plastic Peptide Block Copolymers, Ph.D. thesis, Case Western Reserve University, Cleveland, Ohio, 1976.
24. **Anderson, J. M. et al.,** Solid state structure and mechanical properties of block copolypeptides, in *Advances in Preparation and Characterization of Multiphase Polymer Systems,* Aggarwal, S. L. and Ambrose, R. J., Eds., John Wiley & Sons, New York, 1977, 77.
25. **Uralil, F., Hayashi, T., Anderson, J. M., and Hiltner, A.,** New block copolymers of α-amino acids, *Polym. Eng. Sci.,* 17, 515, 1977.
26. **Hayashi, T., Walton, A. G., and Anderson, J. M.,** Block copolypeptides. I. Synthesis and solid state conformational studies, *Macromolecules,* 10(2), 346, 1977.
27. **Hayashi, T., Anderson, J. M., and Hiltner, P. A.,** Block copolypeptides. II. Viscoelastic properties, *Macromolecules,* 10(2), 352, 1977.
28. **Barenberg, S., Anderson, J. M., and Geil, P. H.,** Structure and properties of two plastic peptide triblock, ABA, copolymers of poly(γ-benzyl-L-glutamate), A, and poly(butadiene/acrylonitrile), B, *Int. J. Biol. Macromol.,* 3, 82, 1981.

29. **Perly, B., Douy, A., and Gallot, B.,** Block copolymers polybutadiene/poly-(benzyl-L-glutamate) and polybutadiene/poly(N⁵-hydroxypropylglutamine) preparation and structural study by x-ray and electron microscopy, *Makromol. Chem.,* 177, 2569, 1976.

30. **Billot, J.-P., Dovy, A., and Gallot, B.,** Synthesis and structural study of block copolymers with a hydrophobic polyvinyl block and a hydrophilic polypeptide block: copolymers polystyrene/poly(L-lysine) and polybutadiene/poly(L-lysine), *Makromol. Chem.,* 117, 1889, 1976.

31. **Douy, A. and Gallot, B.,** Structure of AB block copolymers with a polypeptide block: effect of the chain conformation, *Polym. Eng. Sci.,* 17(8), 523, 1977.

32. **Chen, G.-W., Hayashi, T., and Nakajima, A.,** Synthesis and molecular characterization of A-B-A tri-block copolymers composed of poly(γ-ethyl L-glutamate) as the A component and polybutadiene and the B component, *Polym. J.,* 13(5), 433, 1981.

33. **Hayashi, T., Chen, G.-W., and Nakajima, A.,** Structure and properties of A – B – A tri-block copolymer membranes consisting of poly(γ-benzyl L-glutamate) as the A-component and poly(tetramethyl(eneoxide) as the B-component, *Rep. Prog. Polym. Phys. Jpn.,* 24, 575, 1981.

34. **Chen, G.-W., Sato, H., Hayashi, T., Kugo, K., Noishiki, Y., and Nakajima, A.,** Microheterophase structure, permeability, and biocompatibility of A – B – A tri-block copolymer membranes composed of poly(γ-ethyl L-glutamate) as the A component and polybutadiene as the B component, *Bull. Inst. Chem. Res., Kyoto Univ.,* 59(4), 267, 1981.

35. **Chen, G.-W.,** Microheterophase Structure and some Related Properties of Block Copolymers Containing L- or DL-polyaminoacid as One Component, Ph.D. thesis, Kyoto University, Kyoto, Japan, 1981.

36. **Kugo, K.,** Thermodynamic Studies on Formation of Microheterophase Structures and Formed Surface for Block Copolymers Composed of Polyaminoacid and Polybutadiene, Ph.D. thesis, Kyoto University, Kyoto, Japan, 1982.

37. **Hayashi, T., Kugo, K., and Nakajima, A.,** Structure and physical properties of tri-block copolymer membranes containing poly(γ-methyl L-glutamate) as one component, *Rep. Prog. Polym. Phys. Jpn.,* 25, 687, 1982.

38. **Tabata, Y., Mochizuki, M., Minowa, K., Hayashi, T., and Nakajima, A.,** Synthesis and properties of copolypeptide membranes composed of N-hydroxyethyl glutamine as one component, *Rep. Prog. Polym. Phys. Jpn.,* 25, 683, 1982.

39. **Kugo, K., Hata, Y., Hayashi, T., and Nakajima, A.,** Studies on membrane surfaces of A – B – A tri-block copolymers consisting of poly(ε-N-benzyloxycarbonyl-L-lysine) as the A component and polybutadiene as the B component, *Polym. J.,* 14(5), 401, 1982.

40. **Kugo, K., Murashima, M., Hayashi, T., and Nakajima, A.,** Structure and properties of membrane surfaces of A – B – A tri-block copolymers consisting of poly(γ-methyl D.L-glutamate) as the A component and polybutadiene as the B component, *Polym. J.,* 15(4), 267, 1983.

41. **Kumaki, T. and Imanishi, Y.,** *Polym. Prepr. Jpn.,* 30, 1660, 1981.

42. **Kumaki, T. and Imanishi, Y.,** *Polym. Prepr. Jpn.,* 31, 461, 1982.

43. **Anderson, J. M., Hiltner, A., Schodt, K., and Woods, R.,** Biopolymers as biomaterials: mechanical properties of γ-benzyl-L-glutamate-L-leucine copolymers, *J. Biomed. Mater. Res. Symp.,* 3, 25, 1972.

44. **Anderson, J. M., Gibbons, D. F., Martin, R. L., Hiltner, A., and Wood, R.,** The potential for poly-α-amino acids as biomaterials, *J. Biomed. Mater. Res. Symp.,* 5, 197, 1974.

45. **Hiltner, A., Anderson, J. J., and Borkowski, E.,** Side group motions in poly(α-amino acids), *Macromolecules,* 5(4), 446, 1972.

46. **Sugie, T., Anderson, J. M., and Hiltner, P. A.,** Structure and deformation of a crosslinked poly(α-amino acid), *Polym. Prepr.,* 20, 439, 1979.

47. **Dickinson, H., Sugie, T., Anderson, J. M., and Hiltner, A.,** New poly(α-amino acid) hydrogels for biomedical applications, *Polym. Prepr.,* 20, 39, 1979.

48. **Sugie, T. and Hiltner, A.,** Structure and deformation of a crosslinked poly(α-amino acid), *J. Macromol. Sci. Phys.,* B17(4), 769, 1980.

49. **Sugie, T., Anderson, J. M., and Hiltner, A.,** Mechano-chemical effects in a poly(α-amino acid), *Macromolecules,* 15, 66, 1982.

50. **Oohachi, Y., Hamano, H., Yoshida, T., Tsujita, Y., and Takizema, A.,** Gas permeability and side chain structure of poly(γ-benzyl L-glutamate), *J. Appl. Polym. Sci.,* 22, 1469, 1978.

51. **Kawai, T., Nohmi, T., and Kamide, K.,** Basic studies on membranes for artificial lung, *Polym. Prepr.,* 20, 309, 1979.

52. **Martin, E. C., May, P. D., and McMahon, W. A.,** Amino acid polymers for biomedical applications. I. Permeability properties of L-leucine DL-methionine copolymers, *J. Biomed. Mater. Res.,* 5, 53, 1971.

53. **Klein, E., May, P. D., Smith, J. K., and Leger, N.,** Permeability of synthetic polypeptide membranes. I. Poly-D.L-leucine-co-D.L-methionine, *Biopolymers,* 10, 647, 1971.

54. **Komai, T. and Nose, Y.,** Blood compatibility of glutamate/leucine esters, *Jinko-Zohki,* 4, 114, 1975.

55. **Walton, A. G.,** Synthetic polypeptides as biomedical materials, *Polym. Prepr.,* 20, 294, 1979.

56. **Walton, A. G.**, Interaction of synthetic polypeptides with blood, in *Proceedings of the First Cleveland Symposium on Macromolecules*, Walton, A. G., Ed., Elsevier, Amsterdam, 1977, 286.

57. **Protasi, O., Fabrizi, P., Antoni, G., and Neri, P.**, In-vitro effects on blood coagulation of poly-[$N^5$-(2-hydroxyethyl)-L-glutamine], a synthetic plasma substitute, *Thromb. Res.*, 25, 149, 1982.

58. **Solomon, D. D.**, Interaction of Blood Platelets with Synthetic Copolypeptide Films, Masters thesis, Case Western Reserve University, Cleveland, Ohio, 1977.

59. **Solomon, D. D., Cowin, D. H., Anderson, J. M., and Walton, A. G.**, Platelet interaction with synthetic copolypeptide films, *J. Biomed. Mater. Res.*, 13, 765, 1979.

60. **Barenberg, S. A., Schultz, J. S., Anderson, J. M., and Geil, P. H.**, Hematocompatibility: macromolecular motions and order of the polymer interface, *Trans. Am. Soc. Artif. Intern. Org.*, 25, 159, 1979.

61. **Garcia, C., Anderson, J. M., and Barenberg, S. A.**, Hemocompatibility: effect of structured water, *Trans. Am. Soc. Artif. Intern. Org.*, 26, 294, 1980.

62. **Barenberg, S. A., Anderson, J. M., and Mauritz, K. A.**, Thrombogenesis: an ionic/steric phenomenon, in *Biomaterials 1980*, Winter, G. D., Gibbons, D. F., and Plenk, H., Jr., Eds., John Wiley & Sons, New York, 1982, 451.

63. **Barenberg, S. A., Anderson, J. M., and Mauritz, K. A.**, Thrombogenesis: an epitaxial phenomena. I, *J. Biomed. Mater. Res.*, 15, 231, 1981.

64. **Noishiki, Y., Nakahara, Y., Sato, H., and Nakajima, A.**, Study on the relationship between biocompatibility and chemical structure of amino acid homopolymers and copolymers, *Artif. Org. Jpn.*, 9, 678, 1980.

65. **Marchant, R., Hiltner, A., Hamlin, C., Rabinovitch, A., Slobodkin, R., and Anderson, J. M.**, In vivo biocompatibility studies. I. The cage implant system and a biodegradable hydrogel, *J. Biomed. Mater. Res.*, 17, 301, 1983.

66. **Soderquist, M. E.**, Cellular Interaction with Copolypeptide Films, Masters thesis, Case Western Reserve University, Cleveland, Ohio, 1978.

67. **Walton, A. G., Soderquist, M. E., Solomon, D. D., Cowan, D. E., and Gershman, H.**, Attachment of cells to copolypeptide films, *Polym. Prepr.*, 20, 24, 1979.

68. **Soderquist, M. E., Gershman, H., Anderson, J. M., and Walton, A. G.**, Adhesion of cells to random copolypeptide films, *J. Biomed. Mater. Res.*, 13, 865, 1979.

69. **Sudilovsky, O., Gibbons, D. F., Martin, R. L., and Friedman, L. R.**, In-vitro cytotoxicity and in-vivo acute responses to some amino acid copolymers and their esters, *J. Bioeng.*, 2, 325, 1978.

70. **McKeehan, W. L. and Ham, R. G.**, Stimulation of clonal growth of normal fibroblasts with substrata coated with basic polymers, *J. Cell Biol.*, 71, 727, 1976.

71. **Dickinson, H. R., Hiltner, A., Gibbons, D. F., and Anderson, J. M.**, Biodegradation of a poly($\alpha$-amino acid) hydrogel. I. In-vivo, *J. Biomed. Mater. Res.*, 15, 577, 1981.

72. **Dickinson, H. R. and Hiltner, A.**, Biodegradation of a poly($\alpha$-amino acid) hydrogel. II. In-vitro, *J. Biomed. Mater. Res.*, 15, 591, 1981.

73. **Hayashi, T., Tabata, Y., and Nakajima, A.**, Biodegradation of a poly($\alpha$-amino acid) in vitro. I. Biodegradation of poly(N-hydroxyalkyl L-glutamine) by papain, *Rep. Prog. Polym. Phys. Jpn.*, 26, 587, 1983.

74. **Hayashi, T., Tabata, Y., and Nakajima, A.**, Biodegradation of poly($\alpha$-amino acid) in vitro. II. Biodegradation of copolymers consisting of N-hydroxyalkyl-L-glutamine and L-glutamic acid by papain, *Rep. Prog. Polym. Phys. Jpn.*, 26, 591, 1983.

75. **Marck, K. W., Wildevuur, Ch. R. H., Sederel, W. L., Bantjes, A., and Feijen, J.**, Biodegradability and tissue reaction of random copolymers of L-leucine, L-aspartic acid, and L-aspartic acid esters, *J. Biomed. Mater. Res.*, 11, 405, 1977.

76. **Maurer, P. H.**, Attempts to produce antibodies to a preparation of polyglutamic acid, *Proc. Soc. Exp. Biol. Med.*, 96, 394, 1957.

77. **Maurer, P. H.**, Antigenicity of polypeptides poly($\alpha$-amino acids). II. *J. Immunol.*, 88, 330, 1962.

78. **Maurer, P. H., Gerulat, B. F., and Pinchuck, P.**, Antigenicity of polypeptides (poly alpha amino acids). VII. Studies in humans, *J. Exp. Med.*, 116, 521, 1962.

79. **Maurer, P. H.**, Antigenicity of polypeptides poly($\alpha$-amino acids). XIII. Immunological studies with synthetic polymers containing only D- or D- and L-$\alpha$-amino acids, *J. Exp. Med.*, 121, 339, 1965.

80. **Maurer, P. H.**, Antigenicity of polypeptides poly($\alpha$-amino acids). XVII. Immunologic studies in humans with polymers containing L or D and L-$\alpha$-amino acids, *J. Immunol.*, 95(6), 1095, 1966.

81. **Miyamae, T., Mori, S., and Takeda, Y.**, U.S. Patent 3,371,069, 1968.

82. **Spira, M., Fissette, J., Hall, C. W., Hardy, S. B., and Gerow, F. J.**, Evaluation of synthetic fabrics as artificial skin grafts to experimental burn wounds, *J. Biomed. Mater. Res.*, 3, 213, 1969.

83. **Hall, C. W., Spira, M., Gerow, F., Adams, L., Martin, E., and Hardy, S. B.**, Evaluation of artificial skin models: presentation of three clinical cases, *Trans. Am. Soc. Artif. Intern. Org.*, 16, 12, 1970.

84. **Buckley, C. J., Chambers, C. E., Kemmerer, W. T., Rawling, C. A., Casey, H. W., and Hall, C. W.**, Evaluation of a synthetic bioadherent dressing as a temporary skin substitute, *Trans. Am. Soc. Artif. Intern. Org.*, 17, 416, 1971.

85. **Walder, A. I., May, P. D., Bingham, C. P., and Wright, J. R.,** Evaluation of synthetic films as wound covers, *Trans. Am. Soc. Artif. Intern. Org.,* 15, 29, 1969.

86. **Rowland, G. F., O'Neill, G. J., and Davis, D. A. L.,** Suppression of tumour growth in mice by a drug-antibody conjugate using a novel approach to linkage, *Nature, (London),* 255, 487, 1975.

87. **Shen, W-C. and Ryser, H. J-P.,** Conjugation of poly(L-lysine) to albumin and horseradish peroxidase: a novel method of enhancing the cellular uptake of proteins, *Proc. Natl. Acad. Sci. U.S.A.,* 75(4), 1872, 1978.

88. **Ryser, H. J-P. and Shen, W-C.,** Conjugation of methotrexate to poly(L-lysine) increases drug transport and overcomes drug resistance in cultured cells, *Proc. Natl. Acad. Sci. U.S.A.,* 75(8), 3867, 1978.

89. **Ryser, R. J-P. and Shen, W-C.,** Conjugation of methotrexate to poly(L-lysine) as a potential way to overcome drug resistance, *Cancer,* 45, 1207, 1980.

90. **Petersen, R. V., Anderson, J. M., Fang, S. M., Feijen, J., Gregonis, D. E., and Kim, S. W.,** Studies on norethindrone covalently bonded to poly-N$^5$-(3-hydroxypropyl)-L-glutamine, *Polym. Prepr.,* 20, 20, 1979.

91. **Mitra, S., Van Dress, M., Anderson, J. M., Petersen, R. V., Gregonis, D., and Feijen, J.,** Pro-drug controlled release from polyglutamic acid, *Polym. Prepr.,* 20, 32, 1979.

92. **Zupon, M. A., Christensen, J. M., Petersen, R. V., and Fang, S. M.,** Pharmacokinetics of norethindrone coupled to a biodegradable polymer, *Polym. Prepr.,* 20, 604, 1979.

93. **Gregonis, D., Feijen, J., Anderson, J., and Petersen, R. V.,** A biodegradable contraceptive drug delivery system, *Polym. Prepr.,* 20, 612, 1979.

94. **Petersen, R. V., Anderson, C. G., Fang, S. M., Gregonis, D. E., Kim, S. W., Feijen, J., Anderson, J. M., and Mitra, S.,** Controlled release of progestins from poly(α-amino acid) carriers, in *Controlled Release of Bioactive Materials,* Lewis, D. H., Ed., Academic Press, New York, 1980, 45.

95. **Feijen, J., Gregonis, D., Anderson, C., Petersen, R. V., and Anderson, J.,** Coupling of steroid hormones to biodegradable poly(α-amino acids). I. Norethindrone coupled to poly-N$^5$-(3-hydroxypropyl)-L-glutamine, *J. Pharm. Sci.,* 69(7), 871, 1980.

96. **Petersen, R. V., Fang, S. M., Kim, S. W., Gregonis, D. E., Feijen, J., and Anderson, J. M.,** Biodegradable drug delivery systems based upon poly(L-glutamic acid) and poly(L-glutamines), in *Biomedical Polymers,* Proc. Biol. Eng. Soc. Conf., Biological Engineering Society, U.K., 1982, 211.

97. **Sidman, K. R., Schwope, A. D., Steker, W. D., Rudolph, S. E., and Poulin, S. B.,** Use of synthetic polypeptides for biodegradable drug delivery systems, *Polym. Prepr.,* 20, 27, 1979.

98. **Sidman, K. R., Schwope, A. D., Steker, W. D., Rudolph, S. E., and Poulin, S. B.,** Biodegradable, implantable sustained release systems based on glutamic acid copolymers, *J. Membr. Sci.,* 7, 277, 1980.

99. **Sidman, K. R., Schwope, A. D., Steker, W. D., Rudolph, S. E., Poulin, S. B., and Schnaper, G. R.,** Development and evaluation of a biodegradable drug delivery system, Final Report, for National Institute of Child Health and Human Development, Contract No. N01-HD-4-2802, Arthur D. Little, Cambridge, Mass., 1980.

100. **Sidman, K. R., Schwope, A. D., Steker, W. D., and Rudolph, S. E.,** Use of synthetic polypeptides in the preparation of biodegradable delivery systems for narcotic antagonists, in *Research Monograph 28,* Willette, R. E. and Barnett, G., Eds., National Institute on Drug Abuse, Washington, D.C., 1980, 214.

101. **Sidman, K. R., Steker, W. D., Schwope, A. D., and Schnaper, G. R.,** Controlled release of macromolecules and pharmaceuticals from synthetic polypeptides based on glutamic acid, *Biopolymers,* 22, 547, 1983.

102. **Dohlman, C. H., Pavan-Langston, D., and Rose, J.,** A new ocular insert device for continuous constant-rate delivery of medication to the eye, *Ann. Ophthalmol.,* 4, 823, 1972.

103. **Tani, N., Van Dress, M., and Anderson, J. M.,** Hydrophilic/hydrophobic control of steroid release from a cortisol-polyglutamic acid sustained release system, in *Controlled Release of Pesticides and Pharmaceuticals,* Lewis, D. H., Ed., Plenum Press, New York, 1981, 79.

104. **van Heeswijk, W. A. R., Stoffer, T., Eenink, M. J. D., Potman, W., van der Vijgh, W. J. F., v. d. Poort, J., Pinedo, H. M., Lelieveldl, P., and Feijen, J.,** Synthesis, characterization and antitumor activity of macromolecular prodrugs of adriamycin, in *Recent Advances in Drug Delivery Systems,* Anderson, J. M. and Kim, S. W., Eds., Plenum Press, New York, 1984, 77.

Chapter 5

# SYNTHESES, CHARACTERIZATIONS, AND MEDICAL USES OF THE POLYPENTAPEPTIDE OF ELASTIN AND ITS ANALOGS

**Dan W. Urry and Kari U. Prasad**

## TABLE OF CONTENTS

# I. INTRODUCTION

It has been found by Sandberg and co-workers[1-4] that the precursor protein of fibrous elastin contains repeating peptide sequences. The dominant repeat sequences are a polypentapeptide (PPP) $(L \cdot Val_1 - L \cdot Pro_2 - Gly_3 - L \cdot Val_4 - Gly_5)_n$, a polyhexapeptide (PHP) $(L \cdot Ala_1 - L \cdot Pro_2 - Gly_3 - L \cdot Val_4 - Gly_5 - L \cdot Val_6)_n$, and a polytetrapeptide (PTP) $(L \cdot Val_1 - L \cdot Pro_2 - Gly_3 - Gly_4)_n$. The value of n for the PPP is 11 in pig and 13 in chick (the latter being without a single variation);[5] the value of n for the PHP is greater than 5 for pig; the value of n for the PTP is 4. The monomers, oligomers, and high polymers of these repeats have been synthesized and conformationally characterized.[6] The high polymers have been cross-linked, and the cross-linked PPP and the cross-linked PTP have been found to be elastomeric and capable of an elastic modulus similar to that of the natural elastic fiber.[7-9] The PHP forms cellophane-like sheets and is proposed to assume a precross-linked aligning and interlocking role between protein chains in the fiber.[10,11] Conformational studies using, in addition, cyclic analogs,[12,13] confirmed in the PPP the presence of a recurring secondary structural feature called a ''β-turn'' and have led to a new concept of elasticity referred to as a ''librational entropy mechanism'' of elasticity.[14,15] PPP analogs in which the Gly residues have been replaced by L·Ala and D·Ala residues substantiate the new concept of elasticity and result in additional new biomaterials.[16-18]

Accordingly, the PPP, the most striking primary structural feature of the precursor protein of the elastic fiber, represents a new type of elastomeric biomaterial which is of direct interest and which can be modified in interesting ways. Two intriguing new perspectives have recently become evident. One is that the synthetic elastomeric polypeptide biomaterial can be so constructed as to become covalently cross-linked by tissue enzymes to newly synthesized connective-tissue protein. This is because lysyl oxidase, the natural extracellular cross-linking enzyme, will cross-link PPP chains in which occasional lysine residues have been introduced in position 4.[19] The second new perspective is that the synthetic polypeptide biomaterial can directly be a source of chemotactic peptide for inducing cellular migration into the biomaterial. This derives from the demonstration that the permutation of the hexamer, L·Val–Gly–L·Val–L·Ala–L·Pro–Gly, is a potent chemotactic peptide for elastin-synthesizing fibroblasts.[20] Thus, a synthetic polymer containing the PHP and the PPP in series in a single chain, once cross-linked, and containing occasional lysine residues, would have interesting structural and cellular effector properties for a biomaterial. The biomaterial could induce migration to its matrix of elastin-synthesizing fibroblasts, and it could become chemically cross-linked to the tropoelastin elaborated from the fibroblasts so induced into the material.

Before such perspectives can be realized, however, there are a number of technical problems that require solving; there is the formation of the cross-linked matrix; there are the required characterizations along the way; there are questions of biocompatibility to be addressed; there are perspectives of varied and directed uses of the biomaterial that can be considered. In the present review, the primary problem of synthesizing large quantities of high-molecular-weight PPP at reasonable cost is discussed in terms of a synopsis of the approaches that have been tried, and the present best approach is given. Both chemical and γ-irradiation cross-linking approaches for matrix preparation are presented. Several useful means of verifying the product and characterizing both the high polymer and the cross-linked matrixes are included. Finally, there is reference to tests for biocompatibility and to means of adapting the basic PPP matrix to selected uses.

# II. POLYPENTAPEPTIDE SYNTHESES

Because of its potential as a biomaterial to be desired in large quantity, a number of approaches to the synthesis of the PPP of elastin have been undertaken. One objective has

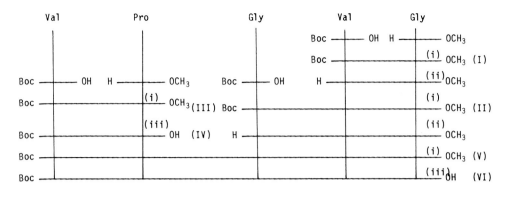

(i) DCC; (ii) HC1/Dioxane: (iii) NaOH

SCHEME 1.   Synthesis of Bocπ Val–Pro–Gly–Val–Gly–OH.

become that of defining a relatively inexpensive synthesis of very high-molecular-weight sequential polypeptide. In what follows are described the strategies and approaches utilized by this laboratory. Also, the approach used by the only other reported synthesis of the PPP of elastin will be noted.[21] The approaches will be presented much as they evolved within the laboratory.

Synthesis of the sequential polypeptides of the repeating pentamer can be achieved by polymerizing any one of the following pentamer permutations:

A.    Val–Pro–Gly–Val–Gly (VPGVG)
B.    Pro–Gly–Val–Gly–Val (PGVGV)
C.    Gly–Val–Gly–Val–Pro (GVGVP)
D.    Val–Gly–Val–Pro–Gly (VGVPG)
E.    Gly–Val–Pro–Gly–Val (GVPGV)

Sequences B and E are of lesser interest as they have a bulky, optically active amino acid at their C-termini which could result in low yields and racemic products. Sequences A and D are of interest since they have the least hindered and optically inactive Gly in the C-terminal position. Sequence C also looks promising as it has the least hindered Gly amino acid at the N-terminus and a Pro at C-terminus which will not result in racemization on activation and coupling of the carboxyl group.

## A. Initial Synthetic Efforts

In the early studies, permutation A was chosen,[22] as it best contains the repeating conformational unit which involves a hydrogen bond between the $Val_1$ C–O and the $Val_4$ NH. The approach is given in Scheme 1 and consists of a strategy of separately synthesizing the Val-Pro dimer and the Gly–Val–Gly trimer, and then coupling these to give the pentamer. This is referred to as a 2 + 3 coupling strategy. The *tert*-butyloxycarbonyl (Boc) group was used for α-amino protection; its removal was achieved with HCl/dioxane, and the coupling reactions were carried out with dicyclohexylcarbodiimide (DCC).[23] The synthesis was then continued to build oligomeric peptides (VPGVG)$_n$ where n = 2 and 3 by the solid phase method of peptide synthesis (SPPS).[24] In this case, the pentamer was first built on the chloromethyl resin with successive couplings of one amino acid at a time followed by the segment couplings using VI obtained by the classical solution methods. The decapeptide and the pentadecapeptide were removed from the resin by HBr/trifluoroacetic acid (TFA) treatment. The peptides were purified by Sephadex column chromatography, formylated,

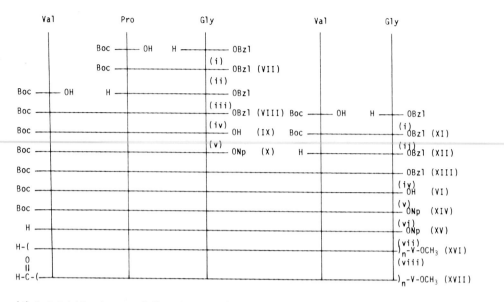

(i) isobutylchloroformate; (ii) HCl/dioxane; (iii) EDCI; (iv) H$_2$/Pd-C; (v) p-nitrophenyl trifluoroacetate; (vi) TFA; (vii) H-Val-OCH$_3$-Et$_3$N/DMSO; (viii) HCO$_2$H/(CH$_3$CO)$_2$O

SCHEME 2.  Synthesis of H$-$$\overset{\text{O}}{\overset{\|}{\text{C}}}$$-$(Val$-$Pro$-$Gly$-$Val$-$Gly)$_n$$-$Val$-$OCH$_3$.

esterified, and completely characterized.[22] The primary purpose of these initial syntheses was to obtain peptides for conformational characterization.

The PPP preparation[25] was carried out by converting VI to the pentachlorophenyl ester,[26] removing the protecting Boc group, and polymerizing the peptide active ester in dimethylformamide. The average value of n in the PPP, H-(VPGVG)$_n$-V-OMe, was approximated to be 10 to 15 by ratioing the formyl *HCO*, and *OCH$_3$* peaks to other peaks in the spectrum such as the Val *CH$_3$* peak. In a second effort at synthesizing the PPP, peptide VI was converted to the *p*-nitrophenyl ester (ON*p*)[27] using *p*-nitrophenol and DCC. The Boc group was removed and the peptide polymerized for 3 days. The result was a PPP with n $\simeq$ 18.[28]

## B. Efforts to Obtain High-Molecular-Weight Polypentapeptide

### 1. Preparation of the Pentamer

Pentamer purity prior to polymerization is a critical factor in obtaining high-molecular-weight polymers in good yield as impurities can result in termination of the polymerization process. In subsequent efforts, great care was taken to obtain very pure active ester, e.g., by passing over a silica gel column and by checking homogeneity using thin-layer chromatography (TLC) with several different solvent systems. Several different approaches were used to facilitate obtaining very pure pentamer. Two approaches are given in Schemes 2 and 3.

In Scheme 2, a 3 + 2 approach was used.[29] Boc-Val-OH was reacted with isobutyl chloroformate[30] to form the mixed anhydride (MA) which was then coupled to the glycine benzyl ester *p*-tosylate and deprotected to give XII. Boc-Pro-OH was coupled to H-Gly-OBzl *p*-tosylate by MA to give VII. The Boc group was removed and coupled to Boc-Val-OH by 1-(3-dimethylaminopropyl)-3-ethylcarbodiimide·HCl (EDCI)[31] to obtain the tripeptide benzyl ester VIII which was hydrogenated in the presence of 10% Pd/C and further converted to *p*-nitrophenyl ester (X) using *p*-nitrophenyl trifluoroacetate.[32] At this stage, the tripeptide was obtained in a crystalline state and therefore could be well purified. X was coupled to XII to obtain the protected pentapeptide benzyl ester XIII. This was hydrogenated to VI, converted to ONp, XIV, deblocked with TFA, and polymerized for 14 days in dimethyl

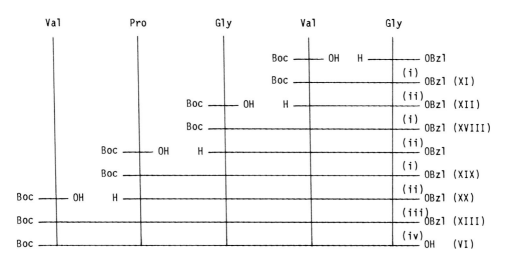

(i) isobutylchloroformate; (ii) HCl/Dioxane; (iii) EDCI; (iv) H₂/Pd-C

SCHEME 3. Synthesis of Boc–Val–Pro–Gly–Val–Gly–OH.

sulfoxide (DMSO) after neutralizing with triethylamine (Et₃N). The reaction was terminated by the addition of H–Val–OCH₃·HCl to give a C-terminal methyl ester.[33] The polymerized material after diluting with water was dialyzed using 3500 mol wt cut-off dialysis tubing and lyophilized. Formylation of XVI was effected with formic acid and acetic anhydride[34] and the PPP, XVII, had n ≥ 40 by end group analysis.

For large-scale preparation of the pentamer, a simple and inexpensive method is desired. For that purpose, MA couplings at each stage of the synthesis is a useful approach. This is schematically shown in Structure 3 for the preparation of the repeat VI.[29] Boc–VG–OBzl (XI) was prepared on a 0.5 $M$ scale with good yields using MA with isobutyl chloroformate and $N$-methyl-morpholine (NMM) at − 15 ± 1°C. After stirring the reaction mixture overnight at room temperature, the solution, on removal of the volatile solvents, was poured into a cold 4% NaHCO₃ solution and stirred for 30 min. The precipitate obtained was filtered, washed with more bicarbonate solution, with water, with 10% citric acid solution, and with water, and then dried to yield TLC pure XI. The synthesis proceeded smoothly to the tetrapeptide, XX, stage. The reaction of Boc-valine with XX by the MA method gave a mixture of the desired product, XIII, and $N$-isobutyloxycarbonyl-PGVG-OBzl in about a 60:40 ratio as determined by proton nuclear magentic resonance spectroscopy. Difficulties of this type, where proline or $N$-methylamino acid was at NH₂-terminal during MA coupling, have also been reported by others.[35-37] At this stage, the coupling step was carried out using EDCI to obtain the desired product XIII in high yield. Recently, this problem with MA coupling was overcome by adding 1-hydroxybenzotriazole to the reaction mixture before the addition of XX. The result was a very good yield of the desired product.[29]

Next, the synthesis of permutation C, GVGVP, is considered. This is the repeat polymerized by Bell et al.[21] using pentachlorophenyl activation of the carboxyl; they reported a high polymer yield of 44% with n = 22. In preparing this pentamer, the 3 + 2 approach was used as shown in Scheme 4. The tripeptide acid, XXI, obtained from XVIII by hydrogenolysis, was coupled with XXII obtained from III by HCl/dioxane treatment using EDCI/HOBt in DMF. Boc–GVGVP–OCH₃ (XXIII) was saponified to XXIV. After having checked the feasibility of preparing the pentamers on a large scale in a pure, simple, and inexpensive way, the next effort is to determine the method of activation for obtaining high yields of very high-molecular-weight polymers.

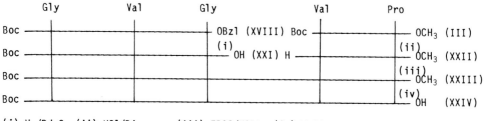

(i) $H_2$/Pd-C; (ii) HCl/Dioxane; (iii) EDCI/HOBt; (iv) NaOH

SCHEME 4.   Synthesis of Boc–Gly–Val–Gly–Val–Pro–OH.

### Table 1
### COMPARISON OF DIFFERENT SOLUTION POLYMERIZATIONS TO OBTAIN HIGH POLYMERS OF THE PPP OF ELASTIN

| No. | Method of synthesis polymerization | Yield (%) after 3500 mol wt cut-off dialysis |
|---|---|---|
| 1 | H–VPGVG–ONp | 70—80 (Several attempts) |
| 2 | H–VPGVG–ONp + 0.2 equivalent HOBt | 78 |
| 3 | H–VPGVG–ONp + 1 equivalent HOBt | 59 |
| 4 | H–VPGVG–ONo | 93 |
| 5 | H–VPGVG–ONm | 74 |
| 6 | H–VPGVG–OPcp | 40 |
| 7 | H–GVGVP–ONp | 94 |
| 8 | H–GVGVP–ONo | 91 |

### 2. Methods of Polymerization

A screening effort was undertaken to determine the activation and polymerization conditions which would give the highest yields of high-molecular-weight PPP. These studies are summarized in Table 1. The yields are given in terms of the quantity of polymer retained in 3500-mol wt cut-off tubing after extensive dialysis against water. A number of syntheses have been carried out using *p*-nitrophenyl ester (ONp) activation and the yields were in the 70 to 80% range but the average number of repeats in the polymer, ñ, was commonly of the order of 40. As hydroxybenzotriazole (HOBt) is often used as a catalyst during active ester couplings both in the classical solution methods and in the solid-phase peptide synthesis methods, it was also used here with the ONp method; 0.2 equivalents gave little improvement of yields whereas 1.0 equivalents resulted in a decrease in yield even with pH monitoring. A similar decrease in yield with increasing HOBt has recently been reported.[38] As Bodanszky et al.[39] have observed *o*-nitrophenyl esters (ONo) to be less sensitive to steric hindrance and solvent effects than ONp, ONo activation was attempted (method 4) and was found to give a high yield, yet the molecular weight remained in the 15,000 range. For purposes of comparison, the *m*-nitrophenyl ester (ONm)[40] method of activation was attempted and found to give a yield comparable to the ONp approach. The pentachlorophenyl ester approach was repeated[25] but the yield remained low (method 6 of Table 1).

Using the GVGVP permutation, both ONp and ONo activations gave high yields (efforts 7 and 8 of Table 1). Moreover, very high molecular weight was obtained with ONp. Further dialysis using 50,000 mol wt cut-off dialysis tubing indicated a 79% yield from the pentamer of greater than 50,000-dalton polymer. Best estimates of ñ are of the order of 200. Thus, a method has been found which gives both high yield and high molecular weights.

## C. Synthesis of Analogs

A series of Ala analogs of the PPP of elastin have been synthesized for the purpose of gaining further insight into the molecular bases for the properties of this interesting PPP and for the purpose of obtaining potential new biomaterials. Using the numbering system for the parent pentamer, L·Val$_1$–L·Pro$_2$–Gly$_3$–L·Val$_4$–Gly$_5$, these analogs can be referred to, e.g., as L·Ala$_5$-PPP, which in single-letter notation would be (VPGVA)$_n$. As a means of indicating the D-configuration as in D·Ala$_3$-PPP, a prime will be added in the single-letter notation, i.e., (VPA′VG)$_n$.

Syntheses of the Ala$_5$-analogs were carried out by means of the pentamer Boc–VXVPG–OBzl. The assembly of the peptide was by stepwise elongation starting from the C-terminal amino acid. The pentapeptide ester was hydrogenated, converted to the ONp, deblocked, and polymerized to obtain L·Ala$_5$PPP[16] and D·Ala$_5$–PPP.[17] Syntheses of the Ala$_3$ analogs were carried out both by the 2 + 3 coupling approach of Scheme 1 and by addition of one amino acid at a time as shown in Scheme 3. Briefly, Boc–VG–OBzl was deblocked and coupled to Boc–D·Ala–OH. The tripeptide so obtained was deblocked and coupled with Boc–VP–OH using EDCI. Alternately, the chain was extended from the deblocked Boc–A′–VG–OBzl by successive couplings with Boc–Pro–OH and Boc–Val–OH. The Boc–VP–A′–VG–OBzl was hydrogenated, converted to ONp, deblocked, and polymerized to obtain the D·Ala$_3$–PPP.[18] The L·Ala$_3$–PPP was similarly synthesized. Finally, the L·Ala$_1$–PPP was synthesized by coupling Boc–APG–OH[41] to H·VG–OCH$_3$ (obtained as shown in Scheme 1) using the MA method to obtain the pentapeptide ester, Boc–APGVG–OCH$_3$. This pentamer was saponified, converted to the ONp, deblocked, and polymerized to obtain the L·Ala$_1$–PPP.[42]

## III. POLYPEPTIDE CHARACTERIZATION

Several aspects of polypeptide characterization are to be considered here: verification of syntheses by nuclear magnetic resonance (NMR) methods, consideration of conformational features derived primarily in solutions using NMR methods but also utilizing X-ray diffraction of related cyclic structures, and discussion of interesting aggregational properties including electron microscopic characterizations.

## A. Verification of Syntheses

In addition to the standard peptide chemistry procedures used above for verifying syntheses, it is possible to demonstrate the purity and amino acid composition of a polypeptide and even to demonstrate independently the sequence of amino acids using carbon-13 and proton NMR approaches. In Figure 1 is the complete carbon-13 NMR spectrum of the PPP. The presence of each carbon atom of the PPP is apparent (see Figure 1B) as is the absence of extraneous peaks. Similarly, in Figure 2 is the proton NMR spectrum of the PPP where again the assignments of all proton resonances and the absence of extraneous peaks, which would arise due to impurities, verify the amino acid composition and purity of the syntheses. As judged from NMR spectra, the issue of purity is answered to within a few percent, i.e., a few percent of racemization, e.g., would not necessarily be detected by these procedures. Additionally, by means of multi-irradiation techniques designed to assess through bond $\alpha CH - NH$ proton-proton coupling interactions and $\alpha CH - C' - O$ and $C' - O - NH$ proton-carbon coupling interactions, as well as to evaluate through space proton-carbon nuclear Overhauser enhancements of $\alpha CH - C' - O$ and $C' - O - NH$, it is possible to walk along the polypeptide backbone atoms and to determine or to verify the sequence of amino acids.[43,44] This has been done for the repeat pentapeptide of elastin and several of its analogs.[16-18]

## B. Conformational Features

Solution NMR studies[6,45] combined with *in vacuo* conformational energy calculations[46]

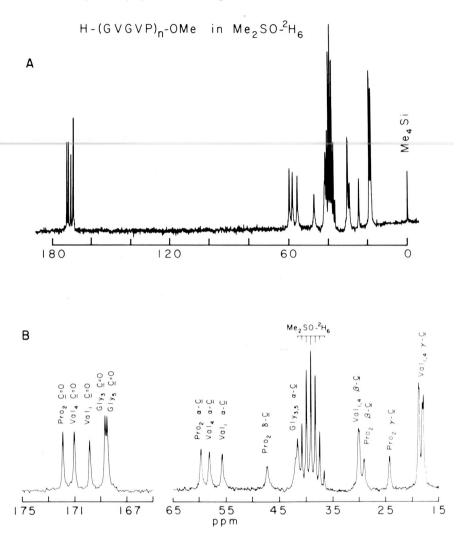

FIGURE 1. Carbon-13 nuclear magnetic resonance (NMR) spectra of the PPP of elastin at 25 MHz in dimethylsulfoxide. (A) Complete spectrum, showing the absence of extraneous peaks; (B) expanded spectra, giving the assignment for every carbon atom in the PPP. This verifies composition and assures purity to within a few percent. The multiple lines near 40 ppm are due to the solvent. The degree of polymerization for the PPP is about 200. (From Urry, D. W., Wood, S. A., Harris, R. D., and Prasad, K. U., *Polymers as Biomaterials,* Shalaby, S. W., Horbett, T., Hoffman, A. S., and Ratner, B., Eds., Plenum Press, New York, in press. With permission.)

as well as X-ray diffraction studies of single crystals of cyclic analogs[13] have demonstrated the presence of the β-turn shown in Figure 3A. The helical recurrence of a β-turn has been termed a β-spiral and the dynamic β-spiral proposed for the PPP of elastin[47] is given as a stereo pair for both axis view (above) and side view (below) in Figure 3B. Also, a schematic representation of the β-spiral is given in Figure 3C.[48] This structure was derived using the concept of cyclic conformations with linear conformational correlates.[12,49,50] The result provides an explanation for the capacity of the PPP to self-align in the formation of structurally anisotropic fibers and yet to be a dominantly entropic elastomeric material (see below). The β-spiral structure has given rise to the librational entropy mechanism of elasticity whereby the entropy derives from the numerous configurations embodied within rocking or torsional oscillations of polypeptide backbone segments.[14,15] This is demonstrated in the most ele-

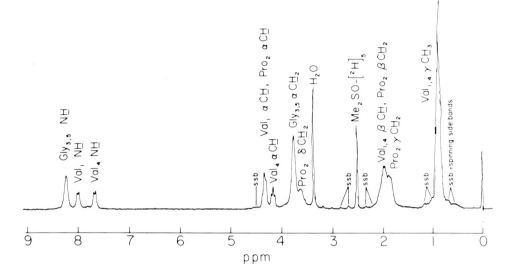

FIGURE 2. Complete proton NMR spectrum taken at 220 MHz in dimethylsulfoxide. The peaks near 3.4 and 2.5 ppm are due to traces of water and due to the solvent. The small humps marked 'ssb' are spinning sidebands of larger peptide and solvent peaks. The assignment of all resonances and the absence of extraneous peaks verifies the composition and purity of the PPP. The number of repeat pentamers in this PPP is about 200, i. e., the polymer has about 1000 residues.

mentary form by the torsional oscillation of the peptide moiety resulting from coupled movements of $\psi_i$ with $\phi_{i+1}$.

## C. Coacervation Properties

The PPP of elastin and the analogs considered here are all soluble in water below 20°C. On raising the temperature toward the physiological range, these polymers associate by means of hydrophobic intermolecular interactions to form cloudy suspensions; on standing the aggregates settle. In the case of the PPP, what forms on the bottom of the vial is a dense viscoelastic phase called a coacervate. The coacervate phase is approximately 50% peptide and 50% water. The process of coacervation is reversible, i.e., on lowering the temperature the coacervate slowly redissolves. Coacervation is a property of tropoelastin, the precursor protein of the elastic fiber[51,52] and of α-elastin, a chemical fragmentation product of the elastic fiber.[53] The temperature profiles for turbidity formation (TPτ) for the high concentration limits of the PPP and a series of analogs are given in Figure 4.[54] By high concentration limit is meant the concentration for which further increases in concentration no longer shift the curves to lower temperatures. It is important and useful to note that the midpoint temperature of the TPτ curve for the high concentration limit also depends on the mean molecular weight. This is shown in Figure 5 where log (molecular weight) is plotted against the midpoint temperature of the TPτ curve. While the data and plot require further refinement, it appears that the TPτ curve itself, for a given polymer, can be used to estimate molecular weight.

## D. Electron Microscopic Characterization

When a drop of turbid solution of curve I of Figure 4 is placed on a carbon-coated grid, negatively stained with 12 m$M$ uranyl acetate-20 m$M$ oxalic acid solution adjusted to pH 6.2 with ammonium hydroxide, and examined in the electron microscope, there is observed a parallel alignment of filaments (see Figure 6).[55] Optical diffraction of the electron micro-

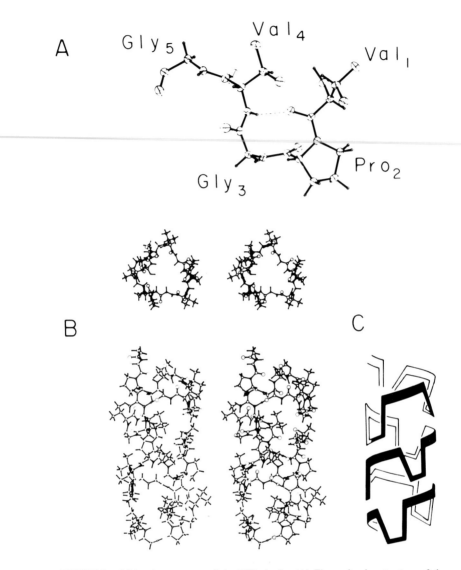

FIGURE 3. Molecular structure of the PPP elastin. (A) The molecular structure of the pentamer repeat in crystalline cyclopentodecapeptide.[13] Note the β-turn evidenced by the Val$_1$ C–O···HN Val$_4$ hydrogen bond;[13] (B) one structure shown as stereopairs of a class of dynamic β-spiral of the PPP of elastin, derived using the approach of cyclic conformations with linear conformational correlates. Note that there is space for water within the β-spiral (seen best in the upper, axis view). The β-turns function as spacers between the turn of the β-spiral and also particularly on slight extension, the β-turns serve to suspend the Val$_4$–Gly$_5$–Val$_1$ peptide segment which is free within the constraints of the β-spiral to occur with numerous variations in the librational states. (From Venkatachalam, C. M. and Urry, D. W., *Macromolecules*, 14, 1225, 1981. With permission.) (C) schematic representation of the dynamic β-spiral of part B. Drawn to show the β-turns functioning as spacers and to show the suspended segments. (From Urry, D. W., *Ultrastruct. Pathol.*, 4, 227, 1983. With permission.) It is the recurring β-turn and, in the relaxed state, hydrophobic interturn interactions that cause the relatively regular β-spiral to form.

graphs shows a number of diffraction spots (see the insert of Figure 6). The most prominent spot, A, occurs at about 55 Å and indicates a lateral spacing for the filaments. This spacing indicates the lateral packing of filaments which, on close examination, have a twisted appearance[10] with a subfilament size that is the approximate diameter of the β-spiral of

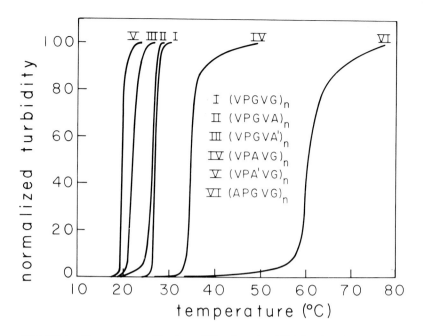

FIGURE 4.  Temperature profiles for turbidity formation (TP$\tau$) for the PPP elastin (curve I) and for the several Ala analogs. See text for discussion. (From Urry, D. W., Walker, J. T., Rapaka, R. S., and Prasad, K. U., *Polym. Preprints*, 24, 3, 1983. With permission.)

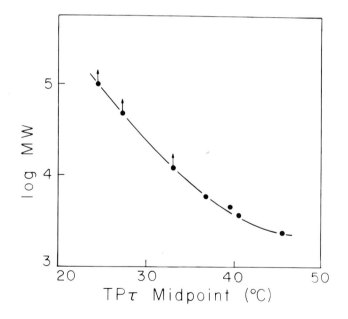

FIGURE 5.  Plot of log (molecular weight) against the midpoint temperature of the TP$\tau$ curves. The data points with arrows pointing to higher molecular weight were fractions retained by a given molecular weight cutoff dialysis membrane and the other points were due to fixed-length peptides synthesized by the solid phase peptide synthesis method with n = 6, 9, 12, and 15.

Figure 3B. The supercoiling of three β-spirals has been shown to give the twisted filament appearance and the approximate dimensions[14] observed in the electron microscopic ultrastructural studies.

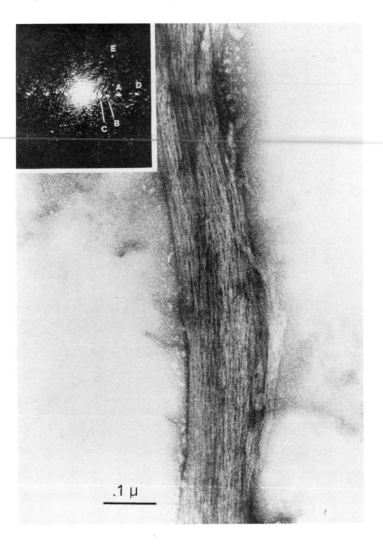

FIGURE 6.   Transmission electron micrograph of negatively stained PPP of elastin obtained from a droplet of cloudy suspension shortly after raising the solution temperature to 40°C. Note the aggregate has a fibrillar shape and is comprised of parallel aligned filaments which on close examination appear to contain twisted subfilaments. Insert: optical diffraction pattern of the electron micrograph showing diffraction spots arising from periodicities in the micrograph. The dominant spot, A, indicates a lateral spacing of about 55 Å for the filament alignment. (From Volpin, D., Urry, D. W., Pasquali-Ronchetti, I., and Gotte, L., *Biochemistry*, 15, 4089, 1976. With permission.)

The heat-aggregated and settled states of the several PPPs are quite different. Even though in Figure 4 the curves for the PPP (I) and the L·Ala₅–PPP (II) fall almost exactly on top of each other, the aggregated states found in the bottom of the vial differ markedly. This is demonstrated in scanning electron micrographs of the dense states that result from heating to 30°C or more. In Figure 7A the PPP coacervate is seen to have a smooth appearance interrupted by cracks that develop on drying. On the other hand, the L·Ala₅–PPP seen in Figure 7B has a granular appearance.[16] Rather than forming a viscoelastic coacervate on warming, the L·Ala₅–PPP forms a granular precipitate. This is the case also for the L·Ala₁–PPP[42] and the L·Ala₃–PPP. The D·Ala₃–PPP, on the other hand, with the level of purity demonstrable by NMR, forms a coacervate phase (see Figure 7C) that is particularly resistant to cracking

A.

B.

C.

FIGURE 7. Scanning electron micrographs of the dense states resulting from heating solutions of PPP and Ala analogs to a temperature above 30°C. (A) The PPP coacervate. This forms a dense viscoelastic phase on heating which on drying exhibits cracks; (B) the L-Ala₅–PPP. The dense phase obtained on warming the sample above 30°C is simply a granular precipitate; (C) the D-Ala₃–PPP coacervate. On warming and standing, this forms a dense viscoelastic phase which is more resistant to cracking on drying. The sample has been scored with a needle to demonstrate the presence of the coacervate. (From Urry, D. W. Trapane, T. L., Wood, S. A., Walker, J. T., Harris, R. D., and Prasad, K. U., *Int. J. Pept. Protein Res.*, 22, 164, 1983. With permission.)

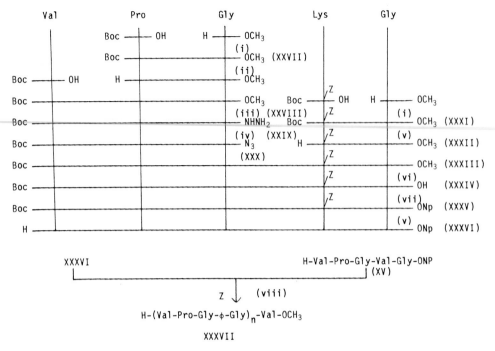

(i) isobutylchloroformate; (ii) HCl/Dioxane; (iii) H₂NNH₂; (iv) isoamylnitrite/HCl; (v) TFA; (vi) NaOH; (vii) p-nitrophenyl trifluoroacetate; (viii) H-Val-OCH₃-Et₃N/DMSO

SCHEME 5.    Synthesis of H–(Val–Pro–Gly–ϕ–Gly)ₙ–Val–OCH₃ (where ϕ = Val or Lys).

on drying.[18] The coacervate phase in Figure 7C has been scored with a needle to demonstrate the presence of the coacervate layer. The D·Ala₅–PPP also develops a coacervate phase but one which requires days rather than hours to form and one which readily cracks on drying.[17] The two D-residue-containing analogs are of particular interest as they offer a potential material that could not be realized by an organism limited in the synthesis of its polypeptide materials to L·amino acids and glycines.

# IV. MATRIX PREPARATION

For most biomaterial roles that might be considered for PPP and its analogs, an insoluble cross-linked matrix is required. The processes of cross-linking are as yet quite primitive but it can be expected in the future that more sophisticated approaches will be practicable. In what follows, chemical and γ-irradiation cross-linking approaches will be briefly considered. Each approach results in a material that is also informative in terms of structure and mechanisms.

## A. Chemical Cross-Linking

Chemical cross-linking can be achieved in innumerable ways. Here it will be demonstrated by the occasional introduction of the trifunctional amino acids lysine (Lys) and glutamic acid (Glu) in place of a Val₄ residue of the PPP. The synthetic approach used is presented in Schemes 5 to 7.[33]

In broad outlines, the approach is to synthesize one PPP in which one out of seven Val₄ residues is replaced by a Lys₄ residue (Scheme 5) and a second PPP in which one out of five Val₄ residues is replaced by a Glu₄ residue. The two PPPs are then to be cross-linked by amide linkage of the Glu γ-carboxyl to the Lys ε-NH₂ group. In more detail, the ε-NH₂

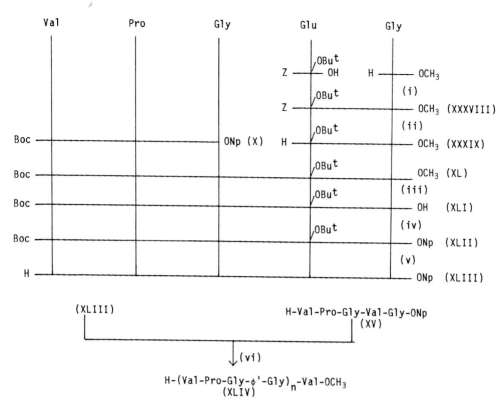

(i) isobutylchloroformate; (ii) H$_2$/Pd-C; (iii) NaOH; (iv) p-nitrophenyl trifluoroace-tate; (v) TFA; (vi) H-Val-OCH$_3$·HCl-Et$_3$N/DMSO

SCHEME 6. Synthesis of H–(Val–Pro–Gly–φ'–Gly)$_n$–Val–OCH$_3$ (where φ' = Val or Glu).

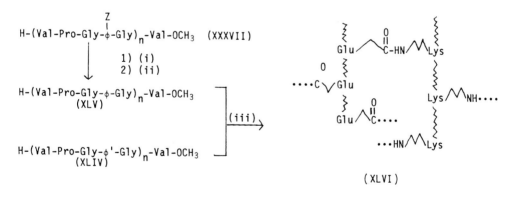

(i) HBr/CH$_3$OH; (ii) Et$_3$N; (iii) CMCI

SCHEME 7. Synthesis of cross-linked polypentapeptide.

group of lysine was protected with benzyloxycarbonyl (Z). The peptides XXVIII and XXXI were synthesized by the MA method. XXVIII was converted to the hydrazide (XXIX) by treatment with hydrazine hydrate which on further treatment with isoamyl nitrite and HCl in cold was transformed into the azide, XXX. The tripeptide azide XXX was reacted with

XXXII obtained from XXXI by the treatment of TFA to yield the pentapeptide methyl ester, XXXIII. Saponification of XXXIII and reaction with *p*-nitrophenyl trifluoroacetate followed by TFA treatment gave XXXVI. This pentapeptide active ester, XXXVI, was copolymerized with H–VPGVG–ONp (XV) in a ratio of 1:7 to obtain the PPP (XXXVII) in which a valine residue in position 4 was replaced by a lysine residue in one out of about seven pentapeptide sequences.

For the syntheses of the Glu-containing pentamer, the removal of Z group from the dipeptide XXXVIII of Scheme 6 was carried out in the presence of pyridine–HCl[56] to minimize the formation of diketopiperazine. The deblocked peptide was coupled to Boc–VPG–ONp(X) to obtain the pentapeptide XL which was saponified to obtain the acid XLI. During the conversion of XLI to the *p*-nitrophenyl ester, a small amount γ-ONp was also formed and the removal of this by-product could be achieved only after treating XLII with TFA and further crystallization. A copolymerization of XLIII with XV in the ratio of 1:4 gave the polypeptide XLIV with n ≥ 40. XLV is derived from XXXVII by HBr/CH₃OH treatment. The chemical cross-linking of XLV and XLIV was then carried out in the presence of 1 − cyclohexyl − 3-(2 − morpholinoethyl) − carbodiimide metho-*p*-toluenesulfonate (CMCI)[57] as outlined in Scheme 7.

The cross-linking of the two polypeptides was achieved both with and without flow orientation. By the flow orientation method, the mixture of PPPs, XLV (100 mg) and XLIV (140 mg), in water (0.7 m𝓁) was first coacervated as a thick coat of film on the sides of a Virtis freeze-drying glass vessel by rotating the latter horizontally, followed by the addition of CMCI and continuing the flowing of the mixture horizontally for several days. The cross-linked peptide XLVI, obtained at 37°C in a flask without flow orientation, was insoluble not only in aqueous solution in which it was synthesized but also in organic solvents, as was the product obtained with flow orientation. The cross-linked material formed without flow orientation was found to have self-assembled into fibers seen in the light microscope without any stain or fixative (see Figure 8A) and seen by scanning electron miroscopy with an aluminum coating to splay-out into fine fibrils and to recoalesce back into the same fiber (see Figures 8B and C).[6] Following the same negative staining procedures as used in Figure 6, the fibrils formed without flow orientation were seen to be comprised of parallel aligned filaments (see Figure 8D)[58] just as seen for the small aggregates found in suspension during coacervation (see Figure 6). This demonstrates quite unambiguously that the PPP is of a molecular structure so constructed as to give rise to anisotropic fibers. This is an important structural conclusion which relates to the mechanism of elasticity (see below).

A chemically cross-linked matrix, formed by flow orientation, is shown in Figure 9 and was suitable for stress-strain studies.[7]

### B. γ-Irradiation Cross-Linking

A simple inexpensive means of obtaining a cross-linked matrix is by γ-irradiation cross-linking using a cobalt-60 source. This becomes a particularly attractive approach once a relatively inexpensive means has been achieved for synthesizing large amounts of very high-molecular-weight PPP as outlined above. Using γ-irradiation cross-linking, the cost to produce a cubic centimeter of matrix is of the order of a few hundred dollars. As the coacervation process is a natural means of forming a uniform and reproducible state, the sample preparation for γ-irradiation cross-linking is simple. The dry PPP can be added to the bottom of a tube; water is added, and the sample is allowed to dissolve in the cold. On raising the temperature to 30°C or above, the coacervate forms in the bottom of the vial and the excess water which is the equilibrium solution above the coacervate can be removed. A plexiglass pestle of appropriate dimension containing a groove or channel is inserted into the tube and the coacervate flows into the circular channel. The tube and pestle containing the shaped vis-coelastic coacervate is then γ-irradiation cross-linked at a particular dose to achieve the

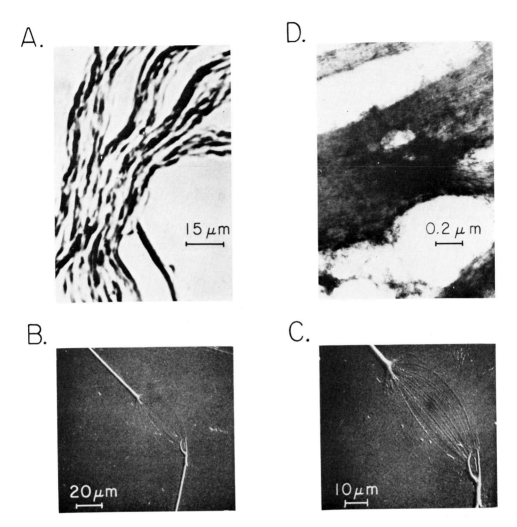

FIGURE 8.   Chemically cross-linked PPP. Weakly cross-linked product XLVI of Scheme 7 which was obtained without flow orientation. The material is seen to have self-organized into fibers and fibrils comprised of filaments. (A) Light micrograph of the material in water showing the formation of fibers with no stain or fixative of any kind present. (From Urry, D. W., *Ultrastruct. Pathol.*, 4, 227, 1983. With permission.) (B and C) Scanning electron micrograph showing a single fiber to splay-out into many fine fibrils which recoalesce to reform the fiber. (From Urry, D. W., Okamoto, K., Harris, R. D., Hendrix, C. F., and Long, M. M., *Biochemistry*, 15, 4083, 1976. With permission.) (D) Transmission electron micrograph of the material negatively stained with uranyl acetate/ oxalic acid shows the fibrils to be comprised of parallel aligned filaments with a periodicity of just over 50 Å. (From Urry, D. W. and Long, M. M., *Elastin and Elastic Tissue*, Vol. 79, Sandberg, L. B., Gray, W. R., and Franzblau, C., Eds., Plenum Press, New York, 1977, 685. With permission.) The PPP of elastin clearly aggregates to form anisotropic fibers due to any underlying filamentous structure which can be related to the β-spiral conformation of Figure 3. See text for further discussion.

desired degree of cross-linking.[59] A cross section of the tube and pestle and a resulting crosslinked matrix are shown in Figure 10.

    The PPP can also be readily compounded to Dacron or some other biomaterial to achieve greater tensile strength and limited extension while retaining the elastomeric properties of the PPP. In previous studies, the DeBakey Elastic Dacron Fabric® (USCI, a division of C. R. Bard, Inc., Cat. No. 007830) has been used.[8,9] The Dacron can be placed in the cut-out channel of the pestle and the pestle with a Dacron strip in the channel can be inserted into

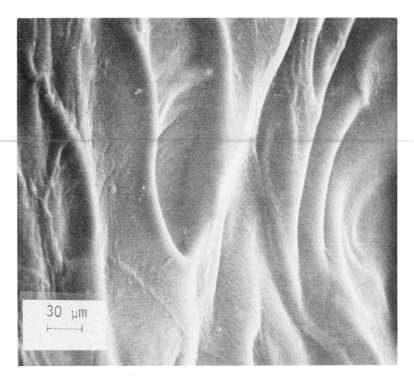

FIGURE 9. Chemically cross-linked PPP of elastin. This is product XLVI of Scheme 7 which was cross-linked during flow orientation to form a wavy matrix seen by scanning electron microscopy. (From Urry, D. W., Okamoto, K., Harris, R. D., Hendrix, C. F., and Long, M. M., *Biochemistry*, 15, 4083, 1976. With permission.)

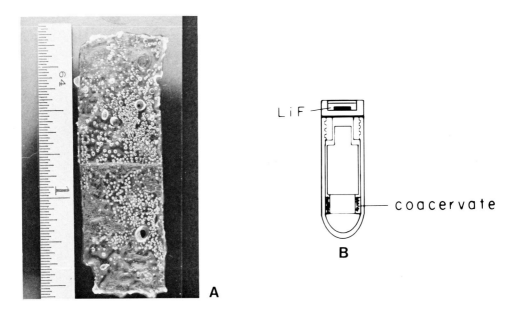

FIGURE 10. (A) Tube containing pestle and coacervate with groove cut in the pestle. The tube is sealed with a cap to prevent drying. Once the coacervate is so shaped it is placed in a cobalt-60 source to be γ-irradiation cross-linked; (B) strip of γ-irradiation cross-linked PPP which is then used for stress-strain, calcification, implant, and other studies.[59]

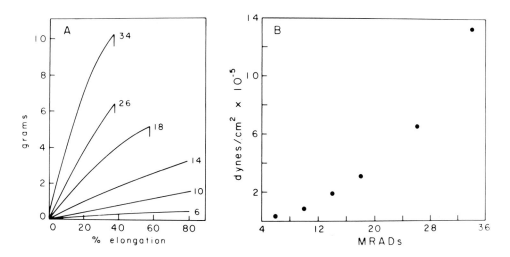

FIGURE 11.    (A) Force-elongation curves, of γ-irradiation cross-linked PPP. The numbers at the right of the curves give the Mrad values. As the γ-irradiation dose increases, the slope increases, indicating an increase in elastic modulus. At 18, 26, and 34 Mrad the elastomeric bands rupture at the indicated elongations; (B) plot of elastic modulus as a function of the MRAD cross-linking dose. A near parabolic relationship is observed. (From Urry, D. W., Wood, S. A., Harris, R. D., and Prasad, K. U., *Polymers as Biomaterials,* Shalaby, S. W., Horbett, T., Hoffman, A. S., and Ratner, B., Eds., Plenum Press, New York, in press. With permission.)

the tube with coacervate in the bottom. The coacervate slowly flows around the pestle, adheres to and permeates through the Dacron, entirely embedding it within the PPP. γ-Irradiation cross-linking both cross-links the PPP to itself and to the Dacron.

## V. ELASTOMERIC PROPERTIES

### A. Stress-Strain Studies

*1. The γ-Irradiation Cross-Linked Polypentapeptide (PPP): $(L \cdot Val_1 - L \cdot Pro_2 - Gly_3 - L \cdot Val_4 - Gly_5)_n$ with $n \simeq 200$*

As shown in Figure 11, the elastic modulus can be controlled by the dose of γ-irradiation. The elastic modulus increases in a near parabolic manner with the increased megaradiation absorbed dose (Mrad). Even with this very high-molecular-weight polymer ($n \simeq 200$), the radiation dose required to achieve higher elastic moduli results in chain damage as evidenced by rupture occurring at ever lower extensions as the Mrad value is increased. While material with low elastic moduli are useful for some applications, for other applications such as vascular prosthetic material, the higher elastic moduli are desirable.

### 2. Compounding to Dacron (a Collagen Analog)

At this stage, it is useful to consider the situation in connective tissue where the elastic and collagen fibers work in concert. As a load is applied, the connective tissue extends with resistance being given by the elastic fiber as it stretches out; the folded collagen fibers unfold until they become extended and come into tension to bear the load at the extension limit. As the load is released, the elastic fibers draw the tissue back to its relaxed configuration. Thus, in developing a prosthetic material for some applications, a collagen analog would be useful. In this vein, Dacron has been used as a collagen analog. As shown in the stress-strain studies of Figure 12, even at 50 Mrad for PPP × Dacron (i.e., PPP compounded to Dacron by γ-irradiation) higher extensions are possible than implied by the data of Figure 11, and an elastic modulus ($\sim 6 \times 10^6$ dyn/cm²) comparable to that of human aorta is obtained.[9] Interestingly, when D·Ala₃–PPP is compounded to Dacron at 30 Mrad, the elastic

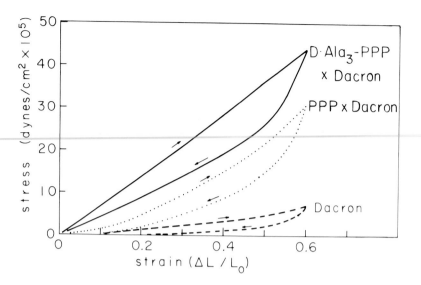

FIGURE 12. Stress-strain curves for DeBakey Elastic Dacron® (— — —), for the same type of Dacron compounded (at 50 Mrad) with the PPP (⋯), and for the D·Ala₃–PPP compounded (at 30 Mrad) to the same Dacron fabric. The hysteresis is due to the Dacron fabric. The elastic modulus for PPP × Dacron is about $6 \times 10^6$ dyn/cm² and that for D·Ala₃–PPP × Dacron is about $10^7$ dyn/cm². See text for discussion. (From Urry, D. W., Trapane, T. L., Wood, S. A., Harris, R. D., Walker, J. T., and Prasad, K. U., *Int. J. Peptide Protein Res.*, 23, 425, 1984. With permission.)

modulus is $\sim 10 \times 10^6$ dyn/cm². This suggests an interesting potential for D·Ala₃–PPP, i.e., $(\text{L·Val}_1-\text{L·Pro}_2-\text{D·Ala}_3-\text{L·Val}_4-\text{Gly}_5)_n$, as a biomaterial. This interesting material, which is not available during the evolution of an organism due to the inability of mammals to incorporate D·residues into polypeptides and proteins, appears capable of achieving a higher elastic modulus at lower irradiation cross-linking doses and also appears to be more cohesive with less cracking on drying (see Figure 7C).

## B. Thermoelasticity Studies

An important characterization of an elastomeric biomaterial involves thermoelasticity studies to determine the fraction of the elastomeric force, f, that is due to internal energy, $f_e$, and that which is due to entropy, $f_s$. For an elastomer to have a long lifetime, it is preferable that it be a perfect entropic elastomer wherein the total elastomeric force derives solely from the decrease on extension of the number of accessible states of essentially identical internal energy. An elastomer with a large $f_e$ component could be expected to have a greater likelihood of chain rupture as the increase in energy on stretching implies stresses being borne by straining the bonds of the polypeptide.

Following the approach applied to elastin by Andrady and Mark[60] which originates in the work of the Flory school,[61] the thermoelasticity studies of Figure 13 were carried out.[59] Plotting ln (f/T) against T allows the ratio of $f_e$:f to be obtained from the slope with knowledge of the coefficient of thermal expansion, β. Since the coefficient of thermal expansion for the PPP coacervate over the 40 to 70°C range is essentially zero, the slope itself gives an estimate of the $f_e$:f ratio. As seen in Figure 13, the slope is very nearly zero. This means that the γ-irradiation cross-linking polypentapeptide is dominantly an entropic elastomer.

Recall that in Figure 8 it was shown that the PPP self-aligns or self-organizes into anisotropic fibers. This requires that the entropic restoring force be achieved in some way with a nonisotropic, nonrandom structure. The proposed librational entropy mechanism derived from the dynamic β-spiral of Figure 3 provides a conformationally based mechanism

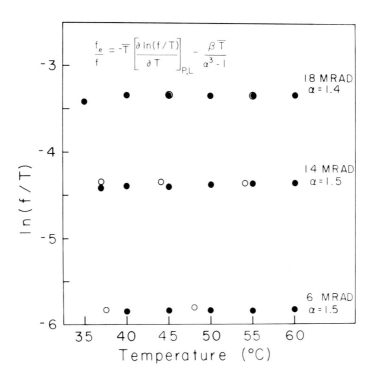

$$\frac{f_e}{f} = -\overline{T}\left[\frac{\partial \ln(f/T)}{\partial T}\right]_{P,L} - \frac{\beta \overline{T}}{\alpha^3 - 1}$$

FIGURE 13. Thermoelasticity studies on elastomeric bands of the PPP cross-linked at the indicated Mrad doses; α gives the elongation with 1.4 being a 40% elongation. The relevant equation is given in the top part of the figure, and indicates that the slope, $\partial \ln(f/T)/\partial T$, gives the value of $f_e/f$ when β can be neglected. The data indicates that $f_e \ll f$ and that the cross-linked PPP is dominantly an entropic elastomer. (From Urry, D. W., Wood, S. A., Harris, R. D., and Prasad, K. U., *Polymers as Biomaterials*, Shalaby, S. W., Horbett, T., Hoffman, A. S., and Ratner, B., Eds., Plenum Press, New York, in press. With permission.)

whereby such elastomers can function. In this structurally based mechanism, the β-turns are seen as suspending the $Val_4$–$Gly_5$–$Val_1$ peptide segment (see Figure 3). As this segment is surrounded by water, it is quite free to undergo changes in its torsion angles. In particular, the motional coupling of the $\psi_4$ and $\phi_5$ and of the $\psi_5$ and $\phi_1$ torsion angles allow for peptide librations, which can readily occur in the suspended segment, to give many states with similar energy. This large number of states which decrease on extension derives to a significant extent from the absence of a side chain in position 5. The $L\cdot Ala_5$–PPP was synthesized to see if replacing an α-hydrogen with a methyl moiety would have any effect on the elastomeric properties. As shown in Figure 7B, this minor substitution entirely destroys the viscoelastic coacervate and results instead in a granular precipitate.[16] While the $D\cdot Ala_5$–PPP will form a coacervate phase on standing for several days, any efforts to stretch the γ-irradiation cross-linked $D\cdot Ala_5$-PPP result in disintegration and no elastomeric retraction.[17] The $D\cdot Ala_3$-PPP, on the other hand, in which the replacement of an α-hydrogen on residue-3 by a methyl stabilizes the β-turn, resulted in a stiffer elastomer (see Figure 12).[18] Thus, the Ala analogs of the PPP support the new perspective for a librational entropy mechanism of elasticity being based on a dynamic β-spiral conformation.

## VI. MEDICAL USES OF THE POLYPENTAPEPTIDE AND ITS ANALOGS

The PPP of elastin $(L\cdot Val_1$–$L\cdot Pro_2$–$Gly_3$–$L\cdot Val_4$–$Gly_5)_n$, has two categories of seemingly disparate potential medical uses. One is the obvious elastomeric prosthetic material with applications ranging from burn cover to vascular wall replacement. The second is a use as a calcifiable matrix with the goal of inducing osteogenesis. This second role was suggested early for Gly-containing sequences of elastin to explain the propensity of the elastic fibers of vascular wall to function as major foci for ectopic calcification.[62] What was proposed was a new mechanism of calcification, referred to as the neutral site binding/charge neutralization mechanism. Gly-containing neutral peptide sequences would bind calcium ion by means of peptide carbonyl oxygens, e.g., initially utilizing the end peptide carbonyl of a β-turn. The neutral matrix would become charged with calcium ions and, as the matrix became positively charged with calcium ions, phosphate ions would be drawn out of solution. The repeat peptides were subsequently identified,[1] and a weakly, chemically cross-linked elastomeric PPP matrix was found to calcify extensively when exposed to normal, preincubated fetal bovine serum dialyzate.[63] The result of exposing a matrix, such as that of Figure 9, to a 12,000-mol wt dialyzate of normal fetal bovine serum is the calcified matrix of Figure 14. This result, buttressed by numerous other studies,[64] constituted demonstration of the proposed new mechanism of calcification.

A major problem of realizing elastomeric medical uses of the PPP of elastin, therefore, would not be unlike the natural process of preserving or insuring the long-term viability of an elastic fiber, and the problem of realizing the calcification and even osteogenic potential of the PPP would be to take advantage of an errant biological process to achieve the desired result. Both would require inducing the desired tissue response to the biomaterial. An intriguing approach here is to utilize chemotatic peptides as briefly noted in the Introduction and as will be considered below. Before doing so, however, the issue of biocompatibility should be addressed.

## A. Biocompatibility

Of course, the first concern when contemplating medical applications of a new material is biocompatibility. At the outset, the situation for the PPP would seem to be promising. The simple aliphatic amino acid composition and the dynamic rather than rigid conformation argue in favor of low antigenicity and low immunogenicity. Importantly, since this is a natural sequence in mammals as diverse as chick and pig and is certainly expected to be present in man, antigens to this repeating sequence are not expected except perhaps in some specific autoimmune disease condition. More directly to the point are studies on the effect of the PPP on two human cell types in tissue culture, human gingival fibroblasts and human skin fibroblasts.[65-67] The pattern of cellular proliferation and surface adhesion properties were unchanged by the presence of PPP as were the cellular and nuclear morphologies. The toxicity assays of direct enumeration and of [51]Cr retention times indicated that the PPP presented no toxicity to the cells.[65-67] Also, cellular metabolism as measured by DNA, general protein, and collagen syntheses were essentially unchanged.[67] In this context, the PPP appears to be quite innocuous. Furthermore, in vivo rabbit implants of γ-irradiation cross-linked PPP, examined radiologically and histologically at 3 and at 13 weeks, presented no foreign body response and no evidence of inflammation.[68]

## B. Applications

Utilization of the PPP as a biomaterial centers on separating the elastomeric material applications from the calcification potential. There are currently three means under consideration for separating the two disparate uses of the PPP. The means, which are such that they may be used in combination, involve the extent of cross-linking of the PPP, the use of analogs, and, a most exciting aspect, the use of chemotactic peptides to induce the appropriate tissue response for the desired application. The more weakly cross-linked the matrix, the more effectively it calcifies when exposed to preincubated fetal bovine serum

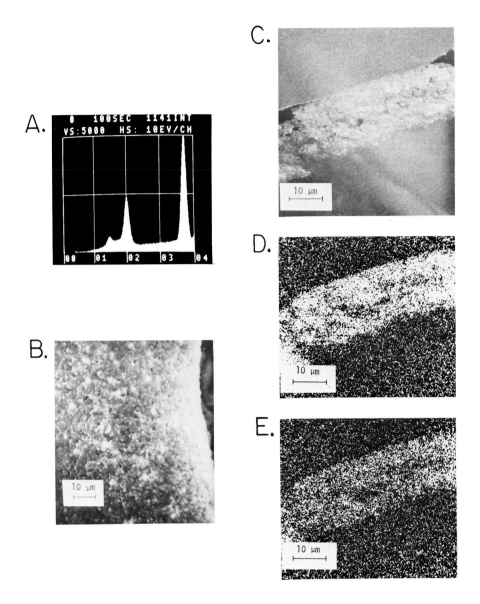

FIGURE 14.   Electron probe microanalysis. (A) Scanning electron micrographs (B and C) and elemental maps (D and E) of weakly chemically cross-linked PPP obtained with flow orientation, as in Figure 9, which has been incubated with normal, preincubated, fetal bovine serum. The X-ray spectrum (A) shows that only the elements P (2.01 keV) and Ca (3.69 keV) were picked up from the serum. The 1.4 keV peak is Al from the coating. The surface of the cross-linked matrix (B) has a rough texture due to calcification. The cross-section of the calcified matrix is seen in C, and the Ca (D) and P (E) maps show that calcification has occurred throughout the matrix. (From Urry, D. W., Long, M. M., Hendrix, and Okamoto, K., *Biochemistry,* 15, 4089, 1976. With permission.)

(FBS). It should be noted, however, that attempts to calcify PPP by exposure to adult human serum have not resulted in a calcified matrix.[85] Preincubation of the serum for 24 hr at 37°C, which is used with FBS, was not effective with adult human serum. This suggests that there may be a more stable inhibitor of calcification in human serum or that nonadult serum has a greater capacity to effect calcification. In either case, the problem of preventing calcification of vascular prosthesis may not be so severe in adults. This is consistent with the finding that replacement heart valves have a more severe problem of calcification in children than in adults.[69-73] With regard to the use of analogs, preliminary calcification studies on γ-irradiation cross-linked D·Ala₃-PPP and D·Ala₅-PPP have suggested that the former analog calcifies less readily than PPP and would be more suitable for vascular prostheses, while the latter analog calcifies more readily than PPP and while not suitable for an elastomeric material (see above), would be a potential calcifiable matrix.[86]

Concerns of medically useful prosthetic biomaterials seem quite inseparable from concerns of wound repair. Quoting from Peacock and van Winkle[74] on the absence of elastin in repair tissue and on the rigidity of scar tissue and its inability to undergo repeated stretching and relaxation, "This failure to include new elastic fibers in the tissue of repair until long after the collagen fibers are formed is another example of the inferiority of scar tissue to the tissue it replaces. It has obvious implications in the repair of skin defects, of ligamentous structures, and of large arteries." A particularly useful elastomeric polypeptide biomaterial, therefore, would be one that would introduce properties deficient in natural repair, perhaps by inducing desirable cellular responses and would be one that could match tissue compliance. Furthermore, it would be particularly favorable if the synthetic biomaterial would become covalently integrated into the newly synthesized connective tissue. Rather than being a foreign body, always discernible, it would be preferable if the biomaterial could be integrated into the newly synthesized fiber components and become essentially indistinguishable from natural connective tissue. This would be a dramatic solution to the problem of graft adhesion.

As noted in the Introduction, there is a repeat hexapeptide sequence in elastin and the L·Val–Gly–L·Val–L·Ala–L·Pro–Gly permutation of the hexamer has been found at concentrations of $10^{-8}$ $M$ to be a potent chemotactic peptide for elastin-synthesizing fibroblasts.[20] Thus, doping the PPP with hexapeptide either by direct inclusion during polymerization or by hexapeptide addition to the coacervate before cross-linking could be expected to induce into the biomaterial the cells that would be required to introduce the elastic component into repair. In addition, if specific lysine containing cross-linking sequences[75-78] or even simply occasional lysines[19] were also included in the biomaterial, then the biomaterial could be expected to become covalently cross-linked to the elastin elaborated by the very cells induced into the biomaterial. Clearly the potential applications for synthetic polypeptide biomaterials modeled after natural sequences is substantial.

One final comment can be made in the use of chemotactic peptides in combination with the PPP matrix. Since this biomaterial has been shown to be a competent calcifiable matrix,[63] its use as such and as a potential substrate for inducing osteogenesis could also be approached by doping with a suitable peptide, one which would be chemotactic toward osteocytes. Such peptides are currently being identified.[79-84] This constitutes an interesting potential application. The general innocuousness of the PPP and its convenient and useful physical properties make it a promising biomaterial for future studies.

## ACKNOWLEDGMENT

This work was supported in part by the National Institutes of Health Grant No. HL-29578 and U.S. Army Medical Research and Development Command Contract No. DAMD17-82-C-2129.

The authors particularly wish to thank M. M. Long for helpful discussions and references, S. A. Wood for assistance in obtaining the matrix of Figure 10, T. L. Trapane for obtaining the data of Figure 2, as well as these and all other past and present members of the Laboratory of Molecular Biophysics who have so significantly contributed to this research program.

# ABBREVIATIONS

| | | | |
|---|---|---|---|
| A | — L-Alanine | NMM | — N-Methyl morpholine |
| A′ | — D-Alanine | NMR | — Nuclear magnetic resonance |
| Boc | — *tert*-Butyloxycarbonyl | | |
| CMCI | — 1–Cyclohexyl–3–(2-morpholinoethyl)–carbodiimide | OBuᵗ | — *tert*-Butyl ester |
| | | OBzl | — Benzyl ester |
| | | $OCH_3$ | — Methyl ester |
| DCC | — Dicyclohexylcarbodiimide | ONm | — *m*-Nitrophenyl ester |
| DMF | — Dimethylformamide | ONo | — *o*-Nitrophenyl ester |
| DMSO | — Dimethylsulfoxide | ONp | — *p*-Nitrophenyl ester |
| EDCI | — 1–(3–Dimethyl aminopropyl)–3–ethyl carbodiimide | OPcp | — Pentachlorophenyl ester |
| | | P | — Proline |
| | | PHP | — Polyhexapeptide |
| $Et_3N$ | — Triethylamine | PPP | — Polypentapeptide |
| G | — Glycine | PTP | — Polytetrapeptide |
| HOBt | — 1-Hydroxybenzotriazole | TFA | — Trifluoroacetic acid |
| MA | — Mixed anhydride | TPτ | — Temperature profile for turbidity formation |
| Mrad | — Mega radiation absorbed dose | | |
| | | V | — Valine |
| | | Z | — Benzyloxycarbonyl |

# REFERENCES

1. **Foster, J. A., Bruenger, E., Gray, W. R., and Sandberg, L. B.,** Isolation and amino acid sequences of tropoelastin peptides, *J. Biol. Chem.*, 248, 2876, 1973.
2. **Sandberg, L. B., Gray, W. R., Foster, J. A., Torres, A. R., Alvarez, V. L., and Janata, J.,** Primary structure of porcine tropoelastin, *Adv. Exp. Med. Biol.*, 79, 277, 1977.
3. **Sandberg, L. B., Soskel, N. T., and Leslie, J. B.,** Elastin structure, biosynthesis and relation to disease states, *N. Engl. J. Med.*, 304, 566, 1981.
4. **Gray, W. R., Sandberg, L. B., and Foster, J. A.,** Molecular model for elastin structure and function, *Nature (London)*, 246, 461, 1973.
5. **Sandberg, L. B.,** private communication.
6. **Urry, D. W. and Long, M. M.,** Conformations of the repeat peptides of elastin in solution: an application of proton and carbon-13 magnetic resonance to the determination of polypeptide secondary structure., *CRC Crit. Rev. Biochem.*, 4, 1, 1976.
7. **Urry, D. W., Okamoto, K., Harris, R. D., Hendrix, C. F., and Long, M. M.,** Synthetic, cross-linked polypentapeptide of tropoelastin: an anisotropic, fibrillar elastomer, *Biochemistry*, 15, 4083, 1976.
8. **Urry, D. W., Harris, R. D., and Long, M. M.,** Irradiation cross-linking of the polytetrapeptide of elastin and compounding to dacron to produce a potential prosthetic material with elasticity and strength, *J. Biomed. Mater. Res.*, 16, 11, 1982.
9. **Urry, D. W. Harris, R. D., and Long, M. M.,** Compounding of elastin polypentapeptide to collagen analogue: a potential elastomeric prosthetic material, *Biomater. Med. Devices Artif. Organs*, 9, 181, 1981.
10. **Urry, D. W.,** Molecular perspectives of vascular wall structure and disease: the elastic component, *Perspect. Biol. Med.*, 21, 265, 1978.
11. **Rapaka, R. S., Okamoto, K., and Urry, D. W.,** Non-elastomeric polypeptide models of elastin: synthesis of polyhexapeptides and a cross-linked polyhexapeptide, *Int. J. Pept. Protein Res.*, 11, 109, 1978.
12. **Urry, D. W., Trapane, T. L., Sugano, H., and Prasad, K. U.,** Sequential polypeptide of elastin: cyclic conformational correlates of the linear polypentapeptide, *J. Am. Chem. Soc.*, 103, 2080, 1981.
13. **Cook, W. J., Einspahr, H. M., Trapane, T. L., Urry, D. W., and Bugg, C. E.,** The crystal structure and conformation of the cyclic trimer of a repeat pentapeptide of elastin, *cyclo*-L–Valyl–L–Prolyl–Glycyl–L–Valyl–Glycl)₃, *J. Am. Chem. Soc.*, 102, 5502, 1980.
14. **Urry, D. W., Venkatachalam, C. M., Long, M. M., and Prasad, K. U.,** Dynamic β-spirals and a librational entropy mechanism of elasticity, in *Conformation in Biology*, G. N. Ramachandran Festschrift Volume, Srinivasan, R. and Sarma, R. H., Eds., Adenine Press, Albany, New York, 1982, 11.

15. **Urry, D. W. and Venkatachalam, C. M.,** A librational entropy mechanism for elastomers with repeating peptide sequences in helical array. *Int. J. Quantum Chem. Quantum Biol. Symp. No. 10*, 81, 1983.

16. **Urry, D. W., Trapane, T. L., Long, M. M., and Prasad, K. U.,** Test of the librational entropy mechanism of elasticity of the polypentapeptide of elastin: effect of introducing a methyl moiety at residue-5, *J. Chem. Soc., Faraday Trans. I*, 79, 853, 1983.

17. **Urry, D. W., Trapane, T. L., Wood, S. A., Walker, J. T., Harris, R. D., and Prasad, K. U.,** D·Ala₅ analog of the elastin polypeptapeptide. Physical characterization, *Int. J. Pept. Protein Res.*, 22, 164, 1983.

18. **Urry, D. W., Trapane, T. L., Wood, S. A., Harris, R. D., Walker, J. T., and Prasad, K. U.,** D·Ala₃ analog of elastin polypentapeptide: an elastomer with an increased Young's modulus, *Int. J. Pept. Protein Res.*, 23, 425, 1984.

19. **Kagan, H. M., Tseng, L., Trackman, P. C., Okamoto, K., Rapaka, R. S., and Urry, D. W.,** Repeat polypeptide models of elastin as substrates for lysyl oxidase, *J. Biol. Chem.*, 255, 3656, 1980.

20. **Senior, R. M., Griffin, G. L., Mecham, R. P., Wrenn, D. S., Prasad, K. U., and Urry, D. W.,** Val–Gly–Val–Ala–Pro–Gly, a repeating peptide in elastin, is chemotactic for fibroblasts and monocytes, *J. Cell. Biol.*, 99, 870, 1984.

21. **Bell, J. R., Boohan, R. C., Jones, J. H., and Moore, R. M.,** Sequential polypeptides. IX. The synthesis of two sequential polypeptide elastin models, *Int. J. Pept. Protein Res.*, 7, 227, 1975.

22. **Urry, D. W., Cunningham, W. D., and Ohnishi, T.,** Studies on the conformation and interactions of elastin. Proton magnetic resonance of repeating pentapeptide, *Biochemistry*, 13(3), 609, 1974.

23. **Sheehan, J. C. and Hess, G. P.,** A new method of forming peptide bonds, *J. Am. Chem. Soc.*, 77, 1067, 1955.

24. **Merrifield, R. B.,** Solid phase peptide syntheses. I. The synthesis of a tetrapeptide, *J. Am. Chem. Soc.*, 85, 2149, 1963.

25. **Urry, D. W., Long, M. M., Cox, B. A., Ohnishi, T., Mitchell, L. W., and Jacobs, M.,** The synthetic polypentapeptide of elastin coacervates and forms filamentous aggregates, *Biochim. Biophys. Acta*, 371, 597, 1974.

26. **Kupryszewski, G.,** Amino acid chlorophenyl esters (II) synthesis of peptide by aminolysis of active N-protected amino acide 2,4,6- trichlorophenyl, esters, *Rocz. Chem.*, 35, 595, 1961.

27. **Bodanszky, M. and Du Vigneaud, V.,** A method of synthesis of long peptide chains using a synthesis of oxytocin as an example, *J. Am. Chem. Soc.*, 81, 5688, 1959.

28. **Urry, D. W., Mitchell, L. W., Ohnishi, T., and Long, M. M.,** Proton and carbon magnetic resonance studies of the synthetic polypentapeptide of elastin, *J. Mol. Biol.*, 96, 101, 1975.

29. **Prasad, K. U., Igbal, M. A., and Urry, D. W.,** Utilization of 1-hydroxy-benzotriazolein in mixed anhydride coupling reactions, *Int. J. Peptide Protein Res.*, 25, 1984.

30. **Vaughan, J. R., Jr., and Osato, R. L.,** The preparation of peptide using mixed carbonic-carboxylic acid anhydrides, *J. Am. Chem. Soc.*, 74, 676, 1952; Anderson, G. W., Zimmerman, J. E., and Callahan, F. M., A reinvestigation of the mixed carbonic anhydride method of peptide synthesis, *J. Am. Chem. Soc.*, 89, 5012, 1967.

31. **Sheehan, J. C., Preston, J., and Cruickshank, P. A.,** A rapid synthesis of oligopeptide derivatives without isolation of intermediates, *J. Am. Chem. Soc.*, 87, 2492, 1965.

32. **Sakakibara, S. and Inukai, N.,** A new reagent for the p-nitrophenylation of carboxylic acids, *Bull. Chem. Soc. Jpn.*, 37, 1231, 1964.

33. **Okamoto, K. and Urry, D. W.,** Synthesis of a cross-linked polypentapeptide of tropoelastin, *Biopolymers*, 15, 2337, 1976.

34. **Stoll, A. and Petrizilka, T.,** Versuche zur syntheses des peptidteils der mutterkornalkaloide I, *Helv. Chim. Acta*, 35, 589, 1952.

35. **Rydon, N. J. and Smith, P. W. G.,** Polypeptides, IV. The self condensation of the esters of some peptides of glycine and proline, *J. Chem. Soc.*, 3642, 1956.

36. **Albertson, N. F.,** Synthesis of peptides with mixed anhydrides, *Org. React.*, 12, 157, 1962.

37. **Beyerman, H. C.,** On the repetitive excess mixed anhydride method for the sequential synthesis of peptides. Synthesis of the sequence of human growth hormone, in *Chemistry and Biology of Peptides*, Meienhofer, J., Ed., Ann Arbor Scientific, Ann Arbor, Mich., 1972, 351.

38. **Yamamoto, H., and Hayakawa, T.,** Synthesis of sequential polypeptides containing L-β-3,4-dehydroxyphenyl-α-alanine (DOPA) and L-lysine, *Biopolymers*, 21, 1137, 1982.

39. **Bodanszky, M., Bath, R. J., Chang, A., Fink, M. L., Funk, K. W., Greenwald, S. M., and Klausner, Y. S.,** Experiments with active esters in solid-phase peptide synthesis, in *Chemistry and Biology of Peptides*, Meinhofer, J., Ed., Ann Arbor Scientific, Ann Arbor, Mich., 1972, 203.

40. **Bodanszky, M.,** Synthesis of peptides by aminolysis of nitrophenyl esters, *Nature (London)*, 175, 685, 1955.

41. **Lorenzi, G. P., Doyle, B. B., and Blout, E. R.,** Synthesis of polypeptides and oligopeptides with the repeating sequence L-alanyl-L-prolylglycine, *Biochemistry*, 10, 3046, 1971.

42. **Rapaka, R. S., Okamoto, K., and Urry, D. W.,** Coacervation properties in sequential polypeptide models of elastin: synthesis of H–(Ala–Pro–Gly–Gly)$_n$–Val–OMe and H–(Ala–Pro–Gly–Val–Gly)$_n$–Val–OMe, *Int. J. Pept. Protein Res.*, 12, 81, 1978.

43. **Khaled, M. A. and Urry, D. W.,** Combined use of multiple and selective proton-decoupled $^{13}$C and $^1$H NMR spectra for complete proton and carbon-13 resonance assignments of polypeptides, *J. Chem. Soc. Chem. Commun.*, 230, 1981.

44. **Khaled, M. A., Harris, R. D., Prasad, K. U., and Urry, D. W.,** Complete proton and carbon-13 resonance assignments of the cyclodecapeptide of elastin by combinative use of multiple and selective proton decoupled $^{13}$C and $^1$H spectra, *J. Magn. Reson.*, 44, 255, 1981.

45. **Urry, D. W.,** Characterization of soluble peptides of elastin by physical techniques, in *Methods in Enzymology*, Vol. 82, Cunningham, L. W. and Frederiksen, D. W., Eds., Academic Press, N.Y., 1982, 673.

46. **Renugopalakrishnan, V., Khaled, M. A., and Urry, D. W.,** Proton magnetic resonance and conformational energy calculations of repeat peptides of tropoelastin: the pentapeptide, *J. Chem. Soc. Perkin Trans 2*, 111, 1978.

47. **Venkatachalam, C. M. and Urry, D. W.,** The development of a linear helical conformation from its cyclic correlate. The β-spiral model of the elastin polypentapeptide, (VPGVG)$_n$, *Macromolecules*, 14, 1225, 1981.

48. **Urry, D. W.,** What is elastin; what is not, *Ultrastruct. Pathol.*, 4, 227, 1983.

49. **Urry, D. W.,** A molecular theory of ion conducting channels: a field dependent transition between conducting and nonconduction conformations, *Proc. Natl. Acad. Sci. U.S.A.*, 69, 1610, 1972.

50. **Urry, D. W.,** Studies on the conformation and interactions of elastin, in *Arterial Mesechyme and Arteriosclerosis*, Vol. 43, Wagner, W. D. and Clarkson, T. B., Eds., Plenum Press, N.Y. 1974, 271.

51. **Smith, D. W., Weissman, N., and Carnes, W. H.,** Cardiovascular studies on copper deficient swine. XII. Partial purification of a soluble protein resembling elastin, *Biochem. Biophys. Res. Commun.*, 31, 309, 1968.

52. **Sandberg, L. B., Weissman, N., and Smith, D. W.,** The purification and partial characterization of a soluble elastin-like protein from copper-deficient porcine aorta, *Biochemistry*, 8, 2940, 1969.

53. **Partridge, S. M., Davis, H. F., and Adair, G. S.,** The chemistry of connective tissues, *Biochem. J.*, 61, 11, 1955.

54. **Urry, D. W., Walker, J. T., Rapaka, R. S., and Prasad, K. U.,** A new class of elastomeric biomaterials: dynamic β-spirals comprised of sequential polypeptides, *Polym. Prep.*, 24, 3, 1983.

55. **Volpin, D., Urry, D. W., Pasquali-Ronchetti, I., and Gotte, L.,** Cross-linked polypentapeptide of tropoelastin: an insoluble, serum calcifiable matrix, *Biochemistry*, 15, 4089, 1976.

56. **Klostermeyr, V.,** Synthese der Teilsequenz 42-49 des Hüllproteins vom Phagen, *Chem. Ber.*, 102, 3617, 1969.

57. **Sheehan, J. C. and Hlavka, J. J.,** The use of water-soluble and basic carbodiimides in peptide synthesis, *J. Org. Chem.*, 21, 439, 1956.

58. **Urry, D. W., and Long, M. M.,** On the conformation, coacervation and function of polymeric models of elastin, in *Elastin and Elastic Tissue*, Vol. 79, Sandberg, L. B., Gray, W. R., and Franzblau, C., Eds., Plenum Press, N.Y., 1977, 685.

59. **Urry, D. W., Wood, S. A., Harris, R. D., and Prasad, K. U.,** Polypentapeptide of elastomeric biomaterial in, *Polymers as Biomaterials*, Shalaby, S. W., Horbett, T., Hoffman, A. S., and Ratner, B., Eds., Plenum Press, New York, in press.

60. **Andrady, A. L. and Mark, J. E.,** Thermoelasticity of swollen elastin networks at constant composition, *Biopolymers*, 29, 849, 1980.

61. **Flory, P. J.,** Molecular interpretation of rubber elasticity, *Rubber Chem. Technol.*, 44, G41, 1968.

62. **Urry, D. W.,** Neutral sites for calcium ion binding to elastin and collagen: A charge neutralization theory for calcification and its relationship to atherosclerosis, *Proc. Natl. Acad. Sci. U.S.A.* 68, 810, 1971.

63. **Urry, D. W., Long, M. M., Hendrix, and Okamoto, K.,** Cross-linked polypentapeptide of tropoelastic: an insoluble, serum calcifiable matrix, *Biochemistry*, 15, 4089, 1976.

64. **Urry, D. W.,** On the molecular basis for vascular calcification, *Perspect. Biol. Med.*, 18, 68, 1974.

65. **Stevens, A., Cogen, R., Urry, D. W., and Long, M. M.,** Effect of a calcifiable matrix on human cell viability, *J. Den. Res.*, 60, 391, 1981.

66. **Waikakul, A., Cogen, R., Stevens, A., Urry, D. W., and Long, M. M.,** Effect of a calcifiable matrix on human cells, *J. Den. Res.*, 61, 189, 1982.

67. **Waikakul, A.,** Effect of a Calcifiable Matrix on Human Cells, M.Sc. thesis, University of Alabama in Birmingham, 1983.

68. **Wood, S. A., Lemons, J. E., Prasad, K. U., and Urry, D. W.,** in preparation.

69. **Sanders, S. P., Levy, R. J., Freed, M. D., Norwood, W. I., and Castaneda, A. R.,** Use of Hancock porcine xenografts in children and adolescents, *Am. J. Cardiol.*, 46, 429, 1980.

70. **Kutsche, L. M., Oyer, P., Shumway, N., and Baum, D.,** An important complication of Hancock mitral valve replacement in children, *Circulation*, 60, Suppl. 1, 98, 1979.

71. **Geha, A. S., Laks, H., Stansel, H. C., Cornhill, J. F., Kilman, J. W., Buckley, M. J., and Roberts, W. C.,** Late failure of porcine valve heterografts in children, *J. Thorac. Cardiovasc. Surg.* 78, 351, 1979.

72. **Bortolotti, U., Milano, A., Mazzucco, A., Galluccci, V., Valente, M., del Maschio, A., Valfre, C., and Thiene, G.,** Alterazioni strurrurali delle bioprotesi di Hancock applicate in eta pediatrica, *G. Ital. Cardiol.,* 10, 1520, 1980.

73. **Williams, W. G., Pollock, J. C., Geiss, D. M., Trusler, G. A., and Fowler, R. S.,** Experience with aortic and mitral valve replacement in children, *J. Thorac. Cardiovasc. Surg.,* 81, 326, 1981.

74. **Peacock, E. E. and Van Winkle, W., Jr., Eds.,** *Wound Repair,* 2nd ed., W. B. Saunders, Philadelphia, 1976.

75. **Sandberg, L. B., Weissman, N., and Gray,** Structural features of tropoelastin related to the sites of cross-links in aortic elastin, *Biochemistry,* 10, 52, 1971.

76. **Foster, J A., Rubin, L., Kagan, H. M., Franzblau, C., Bruenger, E., and Sandberg, L. B.,** Isolation and characterization of cross-linked peptides from elastin, *J. Biol. Chem.,* 249, 6191, 1974.

77. **Gerger, G. E. and Anwar, R. A.,** Comparative studies of the cross-linked regions of elastin from bovine ligamentum nuchae and bovine, porcine and human aorta, *Biochem. J.,* 149, 685, 1975.

78. **Mechan, R. P. and Foster, J. A.,** A structural model for desmosine cross-linked peptides, *Biochem. J.,* 173, 617, 1978.

79. **Malone, J. D., Teitelbaum, S. L., Griffin, G. L., Senior, R. M., and Kahn, A. J.,** Recruitment of osteoclast precursors by purified bone matrix constituents, *J. Cell Biol.,* 92, 227, 1982.

80. **Mundy, G. R. and Poser, J. W.,** Chemotactic activity of the $\gamma$-carboxyglutamic acid containing protein in bone, *Calcif. Tissue Int.,* 35, 164, 1983.

81. **Mundy G. R., Rodan, S. B., Majiska, R. J., DeMartino, S., Trimmier, C., Martin, G. A., and Rodan, G. A.,** Unidirectional migration of osteosarcoma cells with osteoblast characteristics in response to products of bone resorption, *Calcif. Tissue Int.,* 34, 542, 1982.

82. **Orr, F. W., Varani, J., Gondek, M. D., Ward, P. A., and Mundy, G. R.,** Partial characterization of a bone-derived chemotactic factor from tumor cells, *Am. J. Pathol.,* 99, 43, 1980.

83. **Orr, W., Varani, J., Gondek, M. D., Ward, P. A., and Mundy, G. R.,** Chemotactic responses of tumor cells to products of resorbing bone, *Science,* 203, 176, 1979.

84. **Mundy, G. R., Varani, J., Orr, W., Gondek, M. D., and Ward, P. A.,** Resorbing bone is chemotactic for monocytes, *Nature (London),* 275, 132, 1978.

85. **Long, M. M., and Urry, D. W.,** unpublished results.

86. **Woods, and Urry, D. W.,** unpublished results.

Chapter 7

# THE USE AND BIOCOMPATIBILITY OF A HUMAN FIBRIN SEALANT FOR HEMOSTASIS AND TISSUE SEALING

**H. Redl, G. Schlag, and H. P. Dinges**

## TABLE OF CONTENTS

## I. HISTORY AND DEVELOPMENT OF FIBRIN ADHESIVES

The concept of using clotting substances from human blood for hemostasis at internal organs and for wound management traces back to the years of the First World War. In 1915, Grey[1] employed fibrin to control bleeding in operations on the brain, while Harvey[2] used fibrin patches to stop bleeding from abdominal organs during general surgery. In retrospect, the concept appears to have been a perfectly logical one since these substances duplicate the hemostatic process brought about by endogenous substances. However, it took more than 2 decades for this idea to be rediscovered. World War II prompted the development of what in essence is a natural method for hemostasis with the added advantage of an adhesive effect.

In 1940, Young and Medawar[3] reported on experimental nerve anastomosis by fibrin sealing; their intention was to prevent "the disorganization of the fibers which is apt to be produced by stitches." Similarly, Tarlov and Benjamin[4] reunited nerves with plasma clots in 1943. They originally used "fortified cockerel plasma clotted by the addition of chick embryo extract",[5] but later found autologous plasma to be "superior to cockerel plasma since it was unnecessary to augment the fibrin content of the mammalian clot and less inflammatory reaction occurred." Tarlov et al.[6] improved their results obtained by using clotting materials for anastomosis of nerves and avoiding tension at the nerve stump. Our techniques of clot anastomosis of nerves is based on the same principle, i.e., tension at the nerve stump is avoided by interposing nerve grafts.[7] Tarlov et al.[8] even developed a special technique, involving wire supporting rails, to guarantee meticulous apposition of the stumps during the clotting process.

In 1944, Cronkite et al.[9] reported on the first eight cases in which fibrinogen and thrombin had successfully been used for anchoring skin grafts. "The grafts were quickly and well anchored into place, so that fewer sutures or at times none at all were necessary." In the same year, Tidrick and Warner[10] reported on 53 cases in which skin grafting was performed in 122 operations. The authors assumed that, aside from the anchoring effect of fibrinogen plus thrombin, fibrin "also promoted healing". As a result of the poor strength and stability of these seals produced, this method of securing failed to meet expectation and consequently fell into obeyance.

Apparently, the poor adhesivity and stability of the clots in these studies were referable to an inadequate fibrinogen concentration and to absence of antifibrinolytic agents. As a result of the discovery of Factor XIII (fibrin stabilizing factor) in the late 1940s by Laki and Lorand,[11] the properties, actions, and interrelations of fibrin cross-linkage, mechanical strength, and resistance to fibrinolysis were extensively investigated and formed the subjects of numerous publications over the next decade.[12-17]

The use of fibrin as a biological adhesive as we know it today began with the studies of Matras,[18] who successfully employed a fibrinogen cryoprecipitate (produced by the Immuno Co. of Vienna) for reuniting peripheral nerves in animal experiments. It took the development of a special cryoprecipitation process before the production of a highly concentrated fibrinogen solution with an enhanced Factor XIII content became a reality. Additional problems, such as the inhibition of fibrinolysis and the early onset of fibrin degradation, were solved by the introduction of aprotinin,[19] this material being an antiprotease which was first isolated from bovine pancreas by Kunitz and Northrop.[20]

In 1975, Matras and Kuderna[21] reported on the first successful human application of the new fibrin sealant obtained from autologous material for reuniting nerves, an achievement made at Vienna's Lorenz Boehler Traumatology Center.

## II. PHARMACOLOGY

The fibrin sealant is available under the trade names Tissucol®, Tisseel®, or Fibrin-Kleber

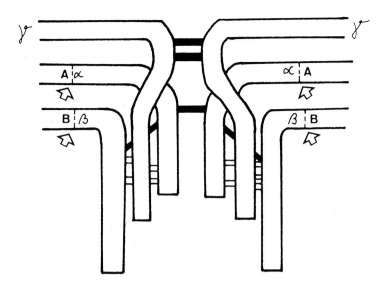

FIGURE 1.    Fibrinogen molecule.

Human Immuno® as a freeze-dried powder in a kit together with thrombin, calcium chloride, and aprotinin solution. The substances in the kit are used to prepare two components: sealer and thrombin solution. To obtain the sealer solution, protein concentrate is dissolved in the accompanying stock solution of fibrinolysis inhibitor (aprotinin 3000 KIU≡mℓ) or a dilution of it, where applicable. Dried thrombin is dissolved in calcium chloride solution to yield the thrombin solution.

Upon reconstitution, 1 mℓ of the sealer solution contains at least 70 mg of fibrinogen, 2 to 7 mg of fibronectin (cold insoluble globulin), 10 units of factor III (1 unit is that amount of Factor XIII which is contained in 1 mℓ of citrated, pooled, human plasma), and 35 μg of plasminogen.

Thrombin is reconstituted in the accompanying calcium chloride solution to yield concentrations of either 500 or 4 US (NIH)-units of thrombin per milliliter, depending on the application method of choice.

The two components are mixed either immediately before application to the recipient surface or in *in situ*. Liquid sealant is a viscous solution with strong adhesive properties. It firmly adheres to the wound surfaces and quickly sets into a white, rubber-like mass, which gains further strength in the course of the next 2 hr. This process is used to achieve hemostasis and to seal or glue wounds or tissue.

## A. Pharmacodynamics

To be active, sealer proteins require the presence of thrombin, calcium chloride, and in some instances, aprotinin solution. Together with these substances, they form a biologic two-component sealant system. As the two components mix after their application, fibrin sealant consolidates and adheres to the site of application, i.e., to the tissue. This explains the usefulness of the sealant for reuniting iatrogenically dissected or traumatically disrupted tissues, sealing wounds, and controlling microvascular bleeding.

The most important of the sealer proteins is fibrinogen, whose molecular weight is about 340,000. The molecule consists of six polypeptide chains of three different types: $\alpha$, $\beta$, and $\gamma$. All chains are connected through disulfide bridges (Figure 1).

By the action of thrombin (Figure 1 — arrows) the fibrinopeptides A and B are split off from the resulting fibrin monomer. These fibrin monomers aggregate mainly because of hydrogen bonding and thus produce the resulting fibrin clot. Like these reactions, the further

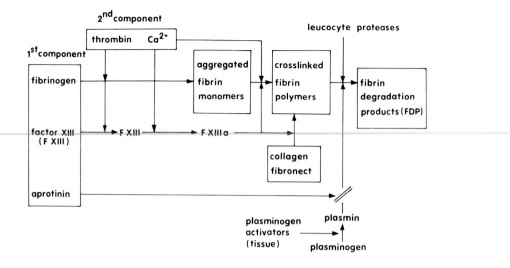

FIGURE 2.   Reaction scheme of the fibrin sealant.

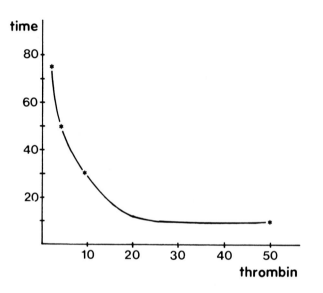

FIGURE 3.   Time dependence of sealant consolidation on the thrombin concentration used at 37°C. For practical reasons we have chosen 4 or 500 units of thrombin, which gives either slow coagulation (30 to 120 sec) or immediate coagulation if mixing is adequate.

consolidation of fibrin sealant duplicates the last phase of the clotting cascade (Figure 2). The time required for the onset of coagulation is dependent on the amount of thrombin used (Figure 3).

Thrombin further activates factor XIII to the transaminase Factor XIIIa. By the action of the latter in presence of Ca ions (we found 5 mmol/$\ell$ absolutely necessary and included 40 mmol/$\ell$ to be on the safer side[22] the fibrin-$\gamma$-chain is immediately (3 to 5 min) cross-linked to the $\gamma$-$\gamma$ dimer. The $\beta$-chain does not react and $\alpha$-chains are cross-linked by peptide-like bridges to an $\alpha$-polymer.

These reactions can be followed by sodium dodecyl sulfate polyacrylamide gel electrophores,[22] which then can be quantitated by densitometry after protein staining (Figure 4). In

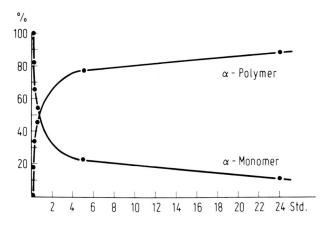

FIGURE 4.    Cross-linking reaction of fibrin α-chains as obtained by densitometric evaluation of polyacrylamide gels. For each gel, the fibrin β-chain band was taken as an internal standard, and the relative decrease of α-monomer was used for the calculation of α-chain polymerization. (From Redl, H., Schlag, G., Dinges, H. P., Kuderna, H., and Seelich, T., *Biomaterials*, Winter, D., Gibbons, D. F., and Plenk, H., Jr., Eds., John Wiley & Sons, New York, 1982. With permission.)

conformity with the results of Mosher,[23] the cross-linking reaction of the fibronectin could also be seen.

As can be seen, the fibrin seal itself contains sufficient factor XIII to produce a high degree of cross-linking. This cross-linking processes slowly, but the initial steepness of the alpha-cross-linkage curve results in sufficient tensile strength after about 3 to 5 min. In previous studies we could demonstrate[23,24] the direct relationship between alpha-chain cross-linking and tensile strength. (Figure 5).

There are also cross-linkages of fibrin with collagen catalyzed by activated factor XIII in the presence of $Ca^{2+}$.[25] In this process plasma fibronectin may play an essential role[26] (Figure 2). In other experiments[27,28] we found that the intrinsic tensile strength of a clot formed with fibrin seal was 1200 g/cm₂ (157 kPa) while that of a sealed rat skin was 200 g/cm₂ (17 kPa) after 10 min cross-linking at 37°C, so that adhesion of the sealant to the tissue is the decisive factor for gluing tissue.

The adhesive qualities of consolidated fibrin sealant to the tissue might be explainable in terms of covalent bonds between fibrin and collagen[26] or fibrin, fibronectin, and collagen. In vitro experiments[12,29,30] have shown that both factor XIII and cross-linked fibrin stimulate fibroblast growth and that the presence of a stabilized fibrin structure (Figure 6) is essential for the formation of a fibroblast network. Fibrinogen did not produce the same effect. Animal experiments[31] corroborate the stimulating effect of factor XIII on wound healing, and they attribute this effect to the growth of fibroblasts.

**B. Pharmacokinetics and Biocompatibility**

In the organism, fibrin sealant undergoes physiologic degradation. The amount of fibrinogen present in fibrin sealant in comparison to total body fibrinogen is low. Assuming that degradation is completed within 24 hr, the daily catabolic rate, even at an extremely high dose (e.g., 10 mℓ or 700 mg of fibrinogen), is thus apt to be no more than 1/30 of that for the endogenous fibrinogen. However, the addition of aprotinin to the fibrin seal rules out this theoretically assumed degradation rate.

Since fibrinolysis is suppressed, only minor fractions of physiologic fibrin degradation are reached, so that intrinsic fibrin catabolism is unaffected. Since the fibrin formed from

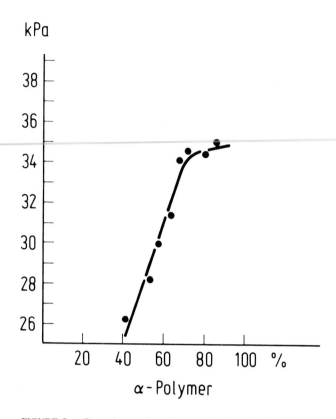

FIGURE 5.    Dependence of tensile strength of standardized fibrin seal clots on the degree of fibrin α-chain polymerization. Clots were prepared by mixing equal volumes of fibrin seal and a solution containing 4 NIH units of thrombin per milliliter and 40 mmol of CaCl2 per liter. For experimental details, see J. Guttmann.[23] (From Redl, H., Schlag, G., Dinges, H. P., Kuderna, H., and Seelich, T., *Biomaterials*, Winter, D., Gibbons, D. F., and Plenk, H., Jr., Eds., John Wiley & Sons, New York, 1982. With permission.)

fibrin sealant application is localized to the site of application (e.g., nerve anastomosis, dural closure), tissue plasminogen activator is likely to be responsible for triggering its degradation. Tissue plasminogen activator binds to fibrin, thus activating fibrin-bound plasminogen (Figure 2). Plasminogen itself is, in fact present in fibrin sealant.

Sealant persistence can, at least to some extent, be controlled by adding an antifibrinolytic agent.[32] Earlier studies showed aprotinin, a natural antiprotease, to be superior to synthetic antifibrinolytic agent,[19] and this was confirmed by others.[33] All the user needs to do, therefore, is to select the proper dosage. It should be remembered in this context that the degradation rate depends on (1) the fibrinolytic (or more generally the proteolytic) activity in the area of application, (2) on the thickness of the sealant layer — which should be as thin as possible, and (3) on the amount of aprotinin present. Thus, problems of dosage can only be dealt with on an individual basis. However, excessively long survival of the sealant is not desirable. It should eventualy yield to wound healing, and one of the remarkable advantages of the sealant is its complete absorbability. While the sealant film should not be resolved before the ingrowth of regenerative tissue, it should not be so thick and persistent as to delay wound healing.

Fibrinolytic activity varies substantially in different tissues. It is, for instance, very high in uterus, moderate in the lungs and kidneys, but much lower in spleen, liver, and bone.

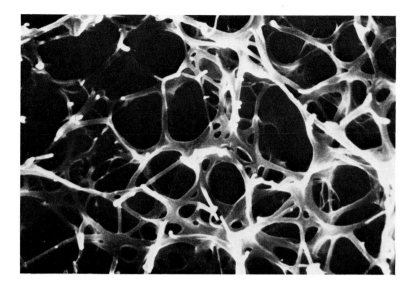

FIGURE 6.    Scanning electron micrograph of a cross-linked fibrin seal clot after freeze-drying.

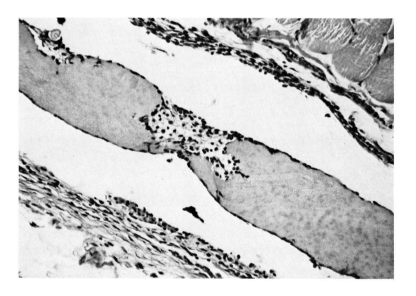

FIGURE 7.    Leukocytes (L) and fibroblasts are penetrating and dissolving a clot of fibrin sealant (FS). (Magnification × 400.)

Animal experiments indicated that tissue degradation (fibrinolysis) was complemented by cellular metabolism or phagocytosis of fibrin sealant by granulocytes and macrophages.[32,34] Granulocytes appear to be prominently involved in the process. In histological specimens they were found to be invariably present in abundance at the fibrin clot margins (Figure 7).

During the first 3 days after sealant application there is only a moderate inflammatory reaction. Several authors,[35-42] found that the most dominating cells in the beginning were neutrophils. According to their work, no histological differences were found between homologous and heterologous sealant for the first 5 days. The neutrophils form a monolayer

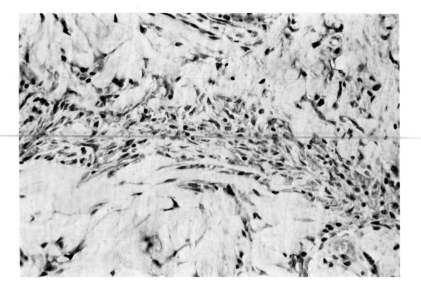

FIGURE 8.   Sealed rat skin. Only fine scar formation (arrow) with little leukocyte infiltration is seen. (Magnification × 400.)

on the fibrin clot, probably as a result of the chemotactic effect of fibrin degradation products or activation of the complement system by plasmin.

Together with the neutrophils, macrophages seem to be responsible for the onset of sealant degradation by release of the the enzymes of the phagocyte. A dense granulocyte infiltration, as was reported to occur for some samples by Albrektsson, [43] is always suspicious of septic processes.[44]

Lymphocytes, plasma cells, and eosinophils are only seen after the 6th day and are significant for an immunological process when heterologous sealant is use. [36,37,45,46]

In most of the publications with histological data, no difference in leukocyte infiltration has been seen between sealant and controls. We obtained the same result in a study [102] in which we sealed lacerations in rat skin with homologous material (Figure 8) where just a few neutrophils could be detected and the finest scars could be obtained. After sealant application (3 to 5 days) a nonspecific granulation tissue develops. Some authors reported foreign giant cells in connection with sealant residues[47,48] which, however, were never seen in our own numerous investigations, either with homologous or heterologous preparations. Results similar to ours have been reported by others[48-53] who definitely mention the lack of foreign-body reaction.

In the final step of wound healing the granulation tissue changes to scar tissue. Scar formation might be dependent on the additives, e.g., factor XIII and inhibitors of fibrinolysis. Kuderns [54] reported fibrosis with high inhibitors of fibrinolysis concentration during peripheral nerve sealing and we[103] could see enhanced scar formation with sealant supplemented with additional factor XIII in experiments on rat skin. One has to keep in mind, however, that while some indication require avoidance of fibrosis, e.g., peripheral nerve sealing, with others this is an even important part of the successful use of fibrin sealant, e.g., dura sealing.

## III. APPLICATION

### A. Reconstitution of the Components

A fibrin sealant kit usually consists of five bottles and all syringes, needles, and a syringe clip (Figure 9). The lyophilized thrombin dissolves quickly upon addition of $CaCl_2$ solution.

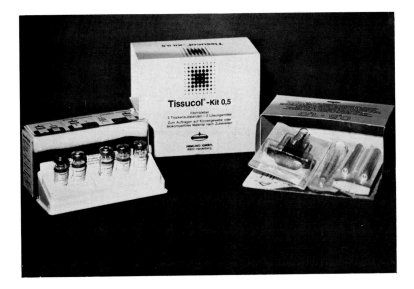

FIGURE 9.    Commercial fibrin sealant kit (Tissucol®) with the freeze-dried material, solutions, and all necessary equipment.

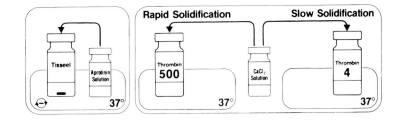

FIGURE 10.    Reconstitution of fibrin sealant. (From the Immuno Co. With permission.)

According to the application, either a low or high thrombin concentration should be used (Figure 10).

To simplify and speed up reconstitution of the highly concentrated sealer proteins we developed a combined heating and stirring device (Figure 11). The preheated aprotinin solution is added to the lyophilysate (Figure 10) and the magnetic stirrer is turned on. (A magnetic bar is found in every sealant vial). Reconstitution is finished within 10 min.

## B. Mixing Components and Delivery to Tissue

Historically the components were always applied sequentially with relatively bad mixing due to fast buildup of fibrin membranes between each other. Like chemical reactions generally, and multi-component adhesives (e.g., epoxy resins) in particular, the sealant system requires an intimate blending of the components for the reaction to be adequate. With few exceptions, [55,56] insufficient attention has been paid to this basic condition, although the major events involved in the reaction were well known. This prompted us to study mixing ratios, and alternative application techniques and their effects on the seal produced. Ever since the first applications of the fibrin sealant, the strength obtainable has been known to depend both on the fibrinogen concentration [18] and on the amount of cross-linkage. [19]

Using a design for measuring intrinsic clot strength, [17] we therefore tried to find an optimum mixing ratio. [28] As expected, clot strength was found to decrease with increasing dilution. There is, however, apparently no point in increasing the fibrinogen concentration

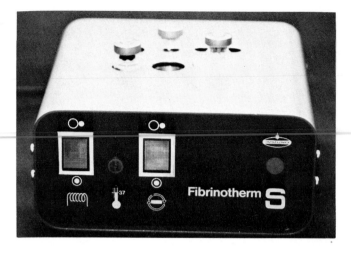

FIGURE 11.   Combined heating and stirring device Fibrinotherm®.

by modifying the sealant to thrombin concentration ratio, as, even under our conditions, the usual mixture of 1 part sealant and 1 part thrombin solution gave the best results. Once again, thorough mixing appears to be the decisive factor.

In a simple test using stained fibrin sealant, inadequate mixing was shown to result in incomplete conversion of the reactants, with the production of an inhomogeneous clot. Some of the clots were found to contain cavities filled with nonconsolidated sealant. Microscopic studies using cancellous bone (Figure 12) showed cavities unevenly filled with sealant upon sequential application.

The gross and microscopic data obtained from experiment on rat skin [28] showed the seals produced with premixed reactants (4 NIH-U thrombin/m$\ell$) or with the Duploject® applicator (4 or 500 NIH-U/m$\ell$ were found to be superior in terms of strength to those obtained with separate application of reactants (Figure 13). No doubt cavitation, as observed microscopically, is one of the factors involved. The insufficient availability of the reactants at the reaction site is another, since adequate cross-linkage requires a minimum concentration of $Ca^{2+}$,[22] which may not be reached locally if mixing is incomplete.

### C. Duploject® System

While we have repeatedly stressed the disadvantages associated with sequential application (poor mixing and cumbersome handling),[19,28] the technique has its place in selected cases, particularly if the sealant film is kept extremely thin, for example by rubbing the sealant into graft material, so that mixing is facilitated. In many cases, application of the sealant components with the double-syringe applicator (Duploject®, Figure 14 and 15) can be expected to offer advantages, e.g., single-handed operation, good mixing, and thin-layer application. As its use is not limited to a specific thrombin concentration, it is almost universally applicable.

Low thrombin concentrations (4 NIH-U/m$\ell$ — slow clotting) are beneficial in all those applications where the parts to be sealed required subsequent adaptation, e.g., in skin grafting and in some microsurgical operations. If, however, hemostasis is of primary interest, high thrombin concentration, i.e., 500 NIH-U/m$\ell$, should be given preference, as they ensure almost instantaneous clotting.

The double-syringe unit with mixing attachment — needle or catheter — is designed for simultaneous operation of the two barrels so that the two components are ejected at the same time but separately via the exchangeable mixing needle. As long as the sealant is being

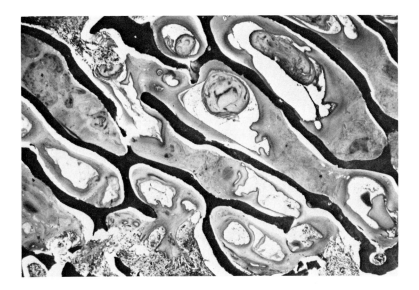

A

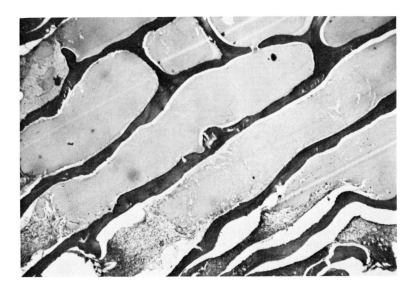

B

FIGURE 12.    Sealant-soaked cancellous bone in the rat skinpouch model. (A) Inhomogeneous fibrin structure after sequential component application; (B) homogeneous sealant
distribution after Duploject® application. (Magnification × 250.)

applied, there will be no clogging of the needle. Once it is interrupted, insertion of a new
needle makes the applicator ready for use again.

## D. Spray Application

The spray head (Figure 16) is connected to sterile gas source such as is commonly used
in operating rooms. The gas pressure is reduced to one to two bar for obtaining a gas flow
of 5 to 10 ℓ/min (Figure 17A and B). The two components are ejected separately into the

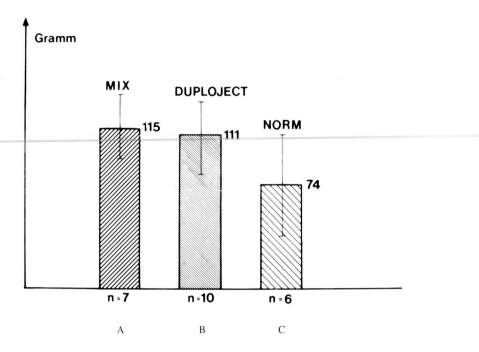

FIGURE 13.   Tensile strength of in vitro sealed rat skin 10 min after sealing (37° C). (A) Premixing technique (4 units of thrombin); (B) Duploject® with mixing cannula (500 units thrombin); (C) application of the two components to different sides (500 units thrombin). (From Redl, H., Schlag, G., and Dinges, H. P., *Thorac. Cardiovasc. Surg.*, 30, 223, 1982. With permission.)

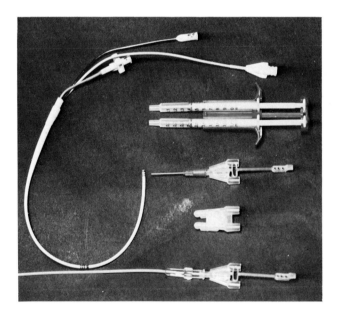

FIGURE 14.   Syringe clip (Duploject®) with different application adaptors.

continuous gas jet. The spray head to wound surface distance is around 10 cm. As the droplets hit each other in the air and on the wound surface they mix, and at a high thrombin concentration, instantly result in a delicate fibrin film. A thin film thus produced is apparently

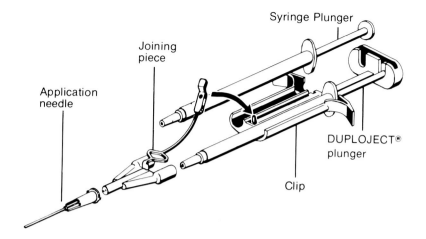

FIGURE 15. Duploject® with mixing needle.

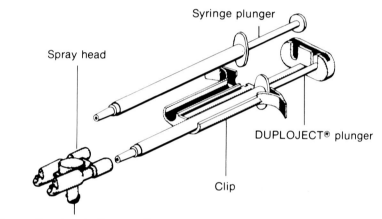

FIGURE 16. Duploject® syringe clip with spray head attachment. Air is supplied from beneath (A) and coming out in the front (B). Components are pressed out (C) into the gas jet. (From the Immuno Co. With permission.)

needed for the sealant to promote wound healing. Spray application also allows the coating of extensive surfaces with a small amount of sealant. Skin grafting and controlling extensive hemorrhage from the liver surface in fully heparinized patients as well as sealing lung tissue are typical examples. An additional advantage offered by the spray head lies in the fact that the gas can be operated separately and the tissue fluid can be blown away by the gas jet. The sealant is thus applied to a dry surface, which improves the adhesion strength of the seal obtained.

For body cavities where it is difficult to use the spray head, one can use a cut Swan-Ganz-Catheter (Figure 18) similar to Linscheer.[58] This technique was successfully used to treat patients with pneumothorax.[59]

For some applications the additional use of sealant support like dacron patches, lyophilized dura, or collagen fleece proved to be useful. However, not all of the commercially available fleece are suitable for this purpose and preliminary tests are therefore required. Some fleeces were tested by Semberger.[33]

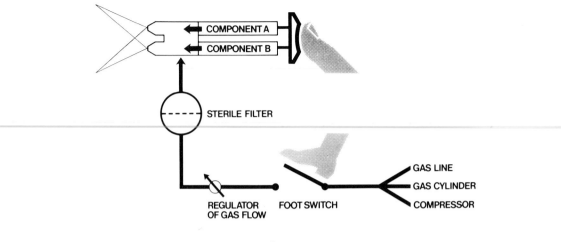

A

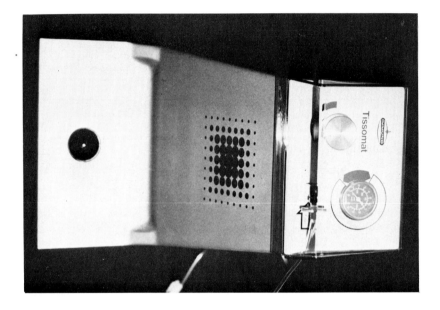

B

FIGURE 17.   (A) Sealant spray applicator: stepping on a foot switch opens the gas line. The gas passes through a sterile filter into the spray head at a constant flow rate. Fluid on the wound surface is blown away from the area to be sealed by the gas jet. As the sealant components are ejected into the gas jet, a sealant film is spread across the wound surface by two overlapping spraying cones (with permission *Thorac. Cardiovasc. Surgeon)*[28]. (B) Tissomat® unit according to A.

We feel that, particularly for Duploject® applications, thin, pliable collagen fleeces will be best suited. Among the few available, a Helitrex® product distributed by American Medical Products has given satisfactory results in terms of wetting, wet tensile strength, and handling.

In summary, for optimum use of the fibrin sealant system, the application technique should meet the requirements:[28]

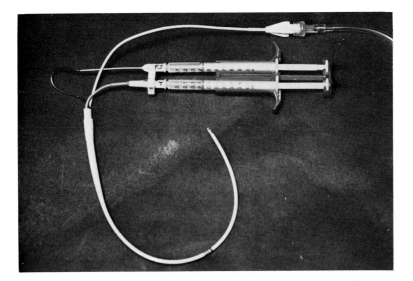

FIGURE 18.    Cut Swan Ganz catheter for spraying proximal lumen—sealer protein (FS) distal lumen to gas line, ballon lumen to thrombin (T).

1.  The sealant components should be fully dissolved and kept at a temperature of 37°C.
2.  The wound surfaces should be as dried as possible.
3.  The component should be thoroughly mixed on application.
4.  The thrombin and aprotinin concentrations should be adjusted to the purpose of application.
5.  The sealant should be applied as a thin film.
6.  After clotting has occurred, further mechanical stresses should be avoided for about 5 min.

### E. Combination of Fibrin Seal with Antibiotics

The practice has been to apply fibrin seal only to area unlikely to become infected. To overcome this limitation, the addition of antibiotics to the fibrin seal seemed desirable. As early as in 1950 a patent was described in the U.S. in which the combined application of fibrin and antibiotics were used.[60] Fibrin seal has also been used in combination with antibiotics both experimentally and clinically.[61,62] Therefore, we studied the in vitro properties of mixtures of fibrin seal and antibiotics. The studies concerned themselves especially with coagulation time, cross-linking, and drug release.[27,63]

For the practical application of fibrin seal, it is important to note that both the clotting time, and the rate of fibrin-α-chain cross-linkage can be regulated by the use of higher thrombin concentrations and additional factor XIII, respectively. Drug release from fibrin seal is probably by simple diffusion, and therefore to a large extent dependent on the concentration gradient between the clot and its environment.

This implies that, although antibiotics incorporated into fibrin clots are retained for a longer period of time than when directly instilled into bone cavities, drug retention is considerable lower than with bone cement-antibiotic mixtures, and not sufficient to maintain high local drug concentrations for more than 3 days. This observation was also confirmed by a recent in vivo study.[64] The limitations might be overcome by new less soluble antibiotics.[65]

Nevertheless, infections might potentially be controlled in the early stages after bone surgery using fibrin seal containing relatively high antibiotic concentrations. However the total dose of drug should be less than the maximal daily systemically recommended dosage.

### F. Detection of Fibrin Seal in Tissues

Due to the opaque white appearance of coagulated fibrin sealant, it is usually easy to detect the fibrin in the sealing area. However, for special indications, e.g., eyes, or with sequential application one might wish to observe the delivery of the sealer protein solution. In these cases, adding disulfine blue dye (ICI) 10 $\mu\ell$/m$\ell$ sealer protein solution is effective in making the fibrin seal visible.

For X-ray detection, the addition of different contrast media were tested by Richling;[66] metrizamide was found to be the more superior but a general use cannot be recommended though there was a slight depression of fibrin-α-chain cross-linking. Histological evaluation of sealed tissues is necessary, especially for experimental work. Reviews on histological techniques for fibrin sealant were published by Dinges[67] and Heine.[68] With the phospho-tungstic acid method by Mallory and the trichrome technique by Lendrum it is possible to visualize the fibrin sealant, but the sealant fibrin does not react as good as endogenous fibrin (perhaps due to the thicker network of fibrin strands). The histological differentiation between the exogenous (sealant) fibrin and the endogenous fibrin needs some personal experience if standard fibrin techniques are employed. If heterologous fibrin glue is used in animal experiments, its demonstration with the immunoperoxidase technique gives optimal results.[67]

## IV. INDICATIONS

Fibrin sealant is indicated wherever it is of interest to reunite traumatically disrupted or surgically dissected tissue, provided the tissue segments to be reunited are not subject to excessive tension. In addition, fibrin sealant has a place in controlling microvascular or capillary bleeding from ruptured or surgically dissected tissues. It is particularly beneficial in patients with increased bleeding tendency undergoing surgery. Fibrin sealant thus appears to have a place in virtually any surgical discipline for purposes of tissue sealing and hemostasis.

### A. Abdominal Surgery

Fibrin sealant was found to be useful in ruptured livers and spleens both for salvaging the organ involved and for sealing cut surfaces after resections. It was used successfully in pancreatic surgery for fistulas and after resections.[69] The use of fibrin sealant has made surgery on the spleen possible.[53,70] In gastric and colonic anastomoses jeoparidized by sutural incompetence, fibrin sealant was found to be beneficial by reinforcing the wound stitches.[52]

### B. Thoracic Surgery

Fibrin sealant was successfully employed for pleural repair in the treatment of idiopathic pneumothorax[59,71] and for airtight seals of lung stitches or clips after partial resection.[72]

### C. Cardiovascular Surgery

In grafting large vessels, fibrin sealant has been used for sealing dacron grafts interposed along the course of the thoracic and abdominal aorta.[73,74]

In cardiovascular surgery, fibrin sealant offers a particular advantage in that it does not require an intact clotting system during extracorporeal circulation. Fibrin sealant can thus be used for "bleeding from stitches, bleeding from coronary anastomosis, fixation of kinked grafts in coronary surgery and diffuse parenchymatous and venous bleeding".[77]

### D. Maxillofacial Surgery

In severe progressive paradontopathies, Schuh[76] used a mixture of homologous bone and fibrin sealant for packing alveolar bone defects and found the grafted bone to be rapidly integrated. Similarly, Matras and Jesch[77] packed epithelial dentogenic cysts with a mix of homologous bone material and fibrin sealant and reported healing to be rapid and uneventful.

## E. Stomatology

In this discipline, fibrin sealant was found to be particularly useful in patients with increased bleeding tendency undergoing tooth extractions.[78-80] Additional preoperative substitution therapy was confined to cases with severe hemophilia.[81]

## F. Neurosurgery

Fibrin sealant has also been employed for extraintracranial anastomoses to remedy occlusions of the internal carotid artery or its branches.[82] In this application, both the number of stitches required and the operating time were reduced. The risk of stenosis was minimized and the duration of ischemia was shortened.

## G. Orthopedics

Mixed with cancellous bone, fibrin sealant has found a place in orthopedics for packing bone cavities. In this application the hemostatic effect of fibrin sealant proved to be particularly beneficial for controlling capillary bleeding into the cavity before introducing the cancellous implant.[83] Similarly, fibrin sealant has given excellent results in controlling hemorrhage from the bone during operations for hemophilic pseudotumors.[84]

## H. Indications in Traumatology

In humans, fibrin sealant was first used for reuniting severed peripheral nerves and for interfascicular nerve grafting. The results obtained after several years of experience were excellent and compared favorably to those seen after perineural suturing.[54]

In fractures of the ankle or comminution of the knee joint, fibrin sealant has been successfully employed without additional osteosynthesis for anchoring osteochondral fragments; it has been found to produce secure bony union.[85] An implant of cancellous material and fibrin sealant gave satisfactory results with uneventful healing[86] when used to pack defects at fracture sites or pseudoarthroses.

## I. Urology

Rauchenwald et al.[87] was one of the first to employ fibrin sealant. He used it successfully for closing nephrotomies performed for removing kidney stones. Similarly, fibrin sealant was employed in pole resections, heminephrectomies, and traumatic kidney rupture.[88,89] After prostatectomies, fibrin sealant was used for its hemostatic effect and was found to significantly reduce the number of blood transfusions required.[90]

## J. Plastic Surgery

Fibrin sealant was used as a hemostatic agent following delayed autologous split-thickness skin grafts in burns and after ablation of necrotic tissue, other means of fixation being omitted.[91] Similarly, the sealant was successfully employed alone without additional stitches to anchor autologous skin grafts in plastic reconstructive surgery of the head and neck region.[92]

## K. Indications of Obstetrics

The sealing effect of fibrin sealant was found to be beneficial in early rupture of the membranes during the second trimester. Amenorrhea was controlled and the fetal membranes were effectively sealed by fibrin sealant.[93]

## L. ENT Indications

In this discipline, Gaspar must be credited with the application of fibrin sealant in tonsillectomies and adenotomies in patients with a history of increased bleeding, particularly in hemophiliacs, where these operations previously carried major risks.[94] Even without

## Table 1
## ADVANTAGES AND DISADVANTAGES OF FIBRIN
## SEALANT ACRYLATES

|  | Fibrin sealant | Acrylates |
|---|---|---|
| Application to wet area | Possible | Impossible |
| Adhesivity | Good | Better |
| Elasticity | Very good | None |
| Tissue compatibility | Excellent | Poor |
| Absorption or degradation | Complete | None |
| Hemostasis | Excellent | None |
| Supporting of wound healing | Obtainable | Unobtainable |
| Application in bone and cartilage | Possible | Impossible |
| Foreign granulation tissue | None | Invariably present |
| Risk of hepatitis | Doubtful | None |

systemic substitution, fibrin sealant alone succeeded in controlling the much dreaded post-operative hemorrhages.

Fibrin sealant also has found applications in the middle ear, the maxillary sinuses, and in fractures of the orbital floor.[95] Its beneficial effect in terms of reinforcing sutures was equally documented in tracheal resections, where a fibrin sleeve was placed around the anastomosal site.[96] In tympanomeatoplasties, fibrin sealant has been employed for anchoring the cartilageous grafts with uneventful healing.[97]

### M. Dermatology

In crural ulcers, conditions for autologous skin grafting are singularly poor because the recipient bed is usually devitalized and infected. Still, Rendl[98] successfully anchored mesh dermatomes at the base of crural ulcers using fibrin sealant and reported a healing rate of 90% In addition, wound healing in skin necroses produced by cytostatic therapy was found to be improved when fibrin sealant was injected under the undermined margins.[99]

## V. ADVANTAGES AND DISADVANTAGES OF FIBRIN SEALANT VS. SYNTHETIC SEALANTS

The major advantage of fibrin sealant lies in its biologic structure and the resultant complete degradation, which is predominantly local. As a consequence, local and systemic toxicity are absent. As fibrin sealant is obtained from human blood derivates, the potential risk of transmitting hepatitis, which has time and again been observed after intravenous fibrinogen application, is a major concern.

Stringent criteria for donor selection such as are employed in obtaining donor plasma for fibrin sealant preparation are apt to reduce the risk of transmitting hepatitis with blood derivatives. However, complete prevention is not yet possible. In two controlled randomized studies, Scheele[100] and Panis[101] did not find the risk of postoperative hepatitis to be increased after fibrin sealant application. Non-A and non-B hepatitis were ruled out by checking transaminase levels. Apparently, the risk of hepatitis is largely reduced by the conversion of fibrinogen to fibrin. The factor or mechanisms underlying viral inactivation or destruction are currently unclear, but the hypothesis appears to be at least legitimate, considering that, among more than 10,000 patients who have so far received fibrin sealant, none developed hepatitis. A comparison between fibrin sealant and synthetic acrylates is given in Table 1.

The only known disadvantage of fibrin sealant is its limited adhesive strength, which will not tolerate major stress exposure. It should, however, be remembered that the objective of using fibrin sealant is not confined to sealing severed tissue segments. Proper adaption of dissociated surfaces is just as important, because it ensures smooth wound healing unhampered by an artificial barrier such as is introduced with synthetic sealants. Fibrin sealant combines hemostasis, tissue sealing, and supporting of wound healing with an optimum physiologic healing mechanism which has given excellent results in almost all surgical disciplines.

## VI. SUMMARY

This history of the fibrin adhesive as a hemostatic agent and its further perfection as a tissue sealant are reviewed and the development of fibrin sealant is highlighted. With the discovery of Factor XIII and its role in the coagulation cascade, the quality of the fibrin sealing technique was substantially improved. Suppression of fibrinolysis eventually made it an optimal system. On the basis of these discoveries, fibrin sealant was introduced into human medicine by Matras.

In the pharmacology section, the pharmacodynamics and pharmacokinetics of the biologic tissue sealant, i.e, the two-component sealant fibrin sealant, are discussed. Its triple effect, hemostasis, tissue sealing, and supporting of wound healing, are well established by numerous animal experiments. The pharmacologic and therapeutic activities of the sealant are briefly presented. Owing to its biologic structure, fibrin sealant is completely degraded and evolves no foreign tissue reaction.

The extensive applications of fibrin sealant in the most varied surgical disciplines are illustrated by numerous examples and the merits of fibrin sealant seals are stressed. Fibrin sealant is compared to synthetic adhesives for its relative advantages and disadvantages. The absence of an increased risk of postoperative hepatitis after fibrin sealant application vs. a control group, such as documented in currently available studies, is emphasized.

Fibrin sealant appears to have gained a firm place in modern surgery. Owing to extensive basic research and the development of a simplified application system it has become an integral tool in many surgical disciplines in combination with bioprosthesis or support materials like collagen fleece.

## REFERENCES

1. **Grey, E. G.,** Fibrin as a hemostatic in cerebral surgery, *Surg. Gynecol. Obstet.,* 21, 452, 1915.
2. **Harvey, S. C.,** The use of fibrin paper and forms in surgery, *Boston Med. Surg.J.,* 658, 1916
3. **Young, J. Z. and Medawar, P. B.,** Fibrin suture of peripheral nerves, *Lancet,* 126, 1940.
4. **Tarlov, I. M. and Benjamin, B.,** Plasma clot and silk suture of nerves, *Surg. Gynecol. Obstet.,* 76, 366, 1943.
5. **Tarlov, I. M. and Benjamin B.,** Autologous plasma clot suture of nerves, *Science,* 95, 1258, 1942.
6. **Tarlov, I. M., Denslow, C., Swarz, S., and Pineles, D.,** Plasma clot suture of nerves, *Arch. Surg.,* 47, 44, 1943.
7. **Kuderna, H.,** Fibrin-Kleber-System—Nervenklebung, *Dtsch. Z. Mund-Kiefer-Gesichtschir.,* 3, 32S, 1979.
8. **Goldfarb, A. I., Tarlov, I. M., Bojar, S., and Wiener, A. S.,** Plasma clot tension strength: its relation to plasma fibrinogen and to certain physical factors, *J. Clin. Invest.,* 22, 183, 1943.
9. **Cronkite, E. P., Lozner, E. L., and Deaver, J. M.,** Use of thrombin and fibrinogen in skin grafting, *JAMA,* 976, 1944.
10. **Tidrick, R. T. and Warner, E. D.,** Fibrin fixation of skin transplants, *Surgery,* 15, 90, 1944.
11. **Laki, K. and Lorand, L.,** On the solubility of fibrin clots, *Science,* 108, 280, 1948.
12. **Beck, E., Duckert, F., Vogel, A., and Ernst, M.,** Der Einfluß des fibrinstabilisierenden Faktors (FSF) auf Funktion und Morphologie von Fibroblasten in vitro, *Z. Zellforsch.,* 57, 327, 1962.

13. **Schwartz, M. L., Pizzo, S. V., Hill, R. L., and McKee, P. A.,** Human factor XIII from plasma and platelets, *J. Biol. Chem.,* 248, 1395, 1973.

14. **Cooke, R. D. and Holbrook, J. J.,** The calcium-induced dissociation of human plasma clotting factor XIII, *Biochem. J.,* 141, 79, 1974.

15. **Curtis, C. G., Brown, K. L., Credo, R. B., Domanik, R. A., Gray, A., Stenberg, P., and Lorand, L.,** Calcium-dependent unmasking of active center cysteine during activation of fibrin stabilizing factor, *Biochemistry,* 13, 3774, 1974.

16. **Shen, L. L., McDonagh, R. P., McDonagh, J., and Hermans, J., Jr.,** Fibrin gel structure: influence of calcium and covalent cross-linking of the elasticity, *Biochem. Biophys. Res. Commun.,* 56, 793, 1974.

17. **Laki, L.,** The biological role of the clot-stabilizing enzymes: transglutaminase and factor XIII, *Ann. N.Y. Acad. Sci.,* 202, 1972.

18. **Matras, H., Dinges, H. P., Lassmann, H., and Mamoli, B.,** Zur nahtlosen interfaszikulären Nerventransplantation im Tierexperiment, *Wr. Med. Wschr.,* 122, 517, 1972.

19. **Redl, H., Schlag, G., Dinges, H. P., Kuderna, H., and Seelich, T.,** Background and methods of fibrin sealing, in *Biomaterials,* Winter, D., Gibbons, D. F., and Plenk, H., Jr., Eds., John Wiley & Sons, New York, 1982, 669.

20. **Kunitz, M. and Northrop, J. H.,** Isolation from beef pancreas of cyrstalline trypsinogen, trypsin, a trypsin inhibitor, and an inhibitor-trypsin compound, *J. Gen.Physiol.,* 19, 991, 1936.

21. **Matras, H. and Kuderna, H.,** Glueing nerve anastomoses with clotting substances, *6th Int. Congr. Plast. & Reconstr. Surg.,* 134, 1975.

22. **Seelich, T. and Redl, H.,** Biochemische Grundlagen zur Klebemethode, *Dtsch. Z. Mund-Kiefer-Gesichtschir.,* 3, 22S, 1979.

23. **Guttmann, J.,** Untersuchung eines Fibrinklebers für die Anwendung, in Der Chirurgie Peripherer Nerven, Diplomarbeit, Technische Universität Wien, 1979.

24. **Seelich, T. and Redl, H.,** Theoretische Grundlagen des Fibrinklebers, in *Fibrinogen, Fibrin und Fibrinkleber,* Schimpf, K., Ed., F. K. Schattauer-Verlag, Stuttgard, 1980, 199.

25. **Mosher, D. F.,** Cross linking of cold insoluble globulin by fibrin stabilizing factor, *J. Biol. Chem.,* 250, 6614, 1975.

26. **Duckert, F., Nyman, D., and Gastpar, H.,** Factor XIII fibrin and collagen, in *Collagen Platelet Interaction,* F. K. Schattauer-Verlag, Stuttgart, 1978, 391.

27. **Redl, H., Stanek, G., Hirschl, A., and Schlag, G.,** Fibrinkleber-Antibiotika-Gemische-Festigkeit und Elutionsverhalten, in *Fibrinkeber in Orthopädie und Traumatologie,* Cotta, H. and Braun, A., Eds., Georg Thieme Verlag, Stuttgart, 1982, 178.

28. **Redl, H., Schlag, G., and Dinges, H. P.,** Methods of fibrin seal application, *Thorac. Cardiovasc. Surg.,* 30, 223, 1982.

29. **Bruhn, H. D., Christopers, E., Pohl, J., and Scholl, G.,** Regulation der Fibroblastenproliferierung durch Fibrinogen/Fibrin, Fibronectin und Faktor XIII, in *Fibrinogen, Fibrin und Fibrinkleber,* Schimpf K., Ed. F. K. Schattauer-Verlag, Stuttgart, 1980, 217.

30. **Turowski, G., Schaadt, M., Barthels, M., Diehl, V., and Poliwada, H.,** Unterschiedlicher Einfluß von Fibrinogen und Faktor XIII auf das Wachstum von Primäar- und Kulturfibroblasten, in *Fibrinogen, Fibrin und Fibrinkleber,* Schimpf, K., Ed., F. K. Schattauer-Verlag, Stuttgart, 1980, 227.

31. **Knoche, H. and Schmitt, G.,** Autoradiographische Untersuchungen Tierexperiment, *Arzneim. Forsch.,* 26, 547, 1976.

32. **Pflüger, H. and Redl, H.,** Abbau von Fibrinkleber in vivo und in vitro (Versuche an der Ratte mit besonderer Berücksichtigung der klinischen Relevanz), *Z. Urol.,* 75, 25, 1982.

33. **Stemberger, A., Fritsche, H. M., Haas, S., Spilker, G., Wriedt-Lübbe, I., and Blümel, G.,** Biochemische und experimentelle Aspekte der Gewebeversiegelung mit Fibrinogen und Kollagen, in *Fibrinkleber in Orthopädie und Traumatoligie,* Cotta, H. and Braun, A., Eds., Georg Thieme Verlag, Stuttgart, 1982, 290.

34. **Pflüger, H., Stocke, W., Kerjaschki, D., and Weissel, M.,** Grundlagen der Nierenparenchymklebung, *Helv. Chir. Acta,* 47, 293, 1980.

35. **Matras H., Jesch, W., Kletter, G., and Dinges, H. P..** Spinale Duraklebung mit ''Fibrinkleber'', *Wr. Klin. Wschr.,* 90. 419, 1978.

36. **Edinger, D., Heine, W. D., Mühling, J., Schröder. F., and Will, Ch.,** Pathohistologie der Wundrandvereinigung mit Fibrinkleber (eine tierexperimentelle Studie), *Dtsch. Z. Mund-, Kiefer-, Gesichts-Chir.,* 4, 166, 1980.

37. **Edinger, D., Heine, W. D., Mühling, J., Schröder, F., and Will Ch.,** Experimental studies on fibrin glue as an adhesive material for skin grafts, *Acta Chir. Maxillo-Facialis,* 8, 120, 1983.

38. **Dinges, H. P., Matras, H., Kletter, G., Chiari, F.,** Histopathologische Untersuchungen zum Heilungsverlauf von gefäßanastomosen bei Anwendung der kombinierten Naht- und Klebetechnik, *VAVA,* 7, 161, 1978.

39. **Dinges, H. P., Redl, H., Kuderna, H., and Matras, H.,** Histologie nach Fibrinklebung, *Dtsch. Z. Mund-Kiefer-Gesichtschir.,* 3, S29, 1979.

40. **Dinges, H. P., Redl, H., Kuderna, H., and Strohmaier, W.,** Histopathologie nach Fibrinklebung, *H. Unfallheilkunde,* 148, 792, 1980.

41. **Pflüger, H., Stackl, W., Kerjaschki, D., and Weissel, M.,** Partial rat kidney resection using autologous fibrinogen thrombin adhesive system, *Urol. Res.,* 9, 105, 1981.

42. **Braun, F., Holle, J., Kova, W., Lindner, A., and Spängler, H. P.,** Untersuchungen über die Replantation autologer Vollhaut mit Hilfe von hochkonzentriertem Fibrinogen and Blutgerinnungsfaktor XIII, *Wr. Klin. Wschr.,* 125, 213, 1975.

43. **Albrektsson, T., Bach, A., Edschage, S., and Jönsson, A.,** Fibrin adhesive system (FAS) influence on bone healing rate (a microradiographical evaluation using the bone growth chamber), *Acta Orthop. Scand.,* 53, 757, 1982.

44. **Frühwald, H. and Dinges, H. P.,** Zum liquordichten Verschluß von Duradefekten, *Laryingol. Rhinol.,* 58, 404, 1979.

45. **Braun, F., Holle, J., Knapp, W., Kovac, W., Passl, R., and Spängler, H. P.,** Immunologische und histologische Untersuchungen bei der Gewebeklebung mit heterologem hochkonzentriertem Fibrinogen, *Wr. Klin. Wschr.,* 87, 815, 1975.

46. **Pflüger, H., Lunglmayr, G., and Breitenecker, G.,** Vaso-Vasoanastomose unter Verwendung von Fibrinogenkonzentraten: Versuche an Kaninchen, *Wr. Klin. Wschr.,* 88, 800, 1976.

47. **Kessler, R., Zimmermann, R. E., Schwering, H., Richter, K. D., and Witting, Ch.,** Anwendung eines Fibrinklebers bei Enteranastomosen, tierexperimentelle Untersuchungen, *Klinische Ergebnisse,* in press.

48. **Zilch, H. and Friedebold, G.,** Klebung osteochondraler Fragmente mit dem Fibrinkleber — Klinische Erfahrungen, *Akt. Traumatol.,* 11, 136, 1981.

49. **Spilker, G., Fischer, M., Stemberger, A., Fritsche, H. J., Meierhofer, J. N., Haas, S., and Blümel, G.,** Fibrinklebung nach ausgedehnten Leberparenchymdefekten - ein neues therapeutsiches Verfahren in der Leberchirurgie, in *Experimentelle und Klinische Hepatologie,* Vol. 2, *Arbeitstagung Experimentelle und Klinische Hepatologie,* Zelder, O., Fischer, M., Eckert, P., and Bode, J. C., Eds., Thieme-Verlag, Stuttgart, l980, 25.

50. **Heidecker, C. D.,** Experimentelle Untersuchungen von Enterotomien des Rattenileums nach Naht und Applikation von physiologischen Plasmafraktionen, Dissertation, München, 1980.

51. **Spängler, H. P.,** Gewebeklebung und lokale Blutstillung mit Fibrinogen, Thrombin und Blutgerinnungsfaktor XIII (Experimentelle Untersuchungen und klinische Erfahrungen, *Wr. Klin, Wschr.,* 88(49), 1976.

52. **Scheele, J., Herzog, J., and Mühe, E.,** Anastomosensicherung am Verdauungstrakt mit Fibrinkleber, Nahttechnische Grundlagen, experimentelle Befunde, Klinische Erfahrungen, *Zbl. Chir.,* 103, 1325, 1978.

53. **Brands, W., Beck, M., and Raute-Kreinsen, U.,** Gewebeklebung der rupturierten Milz mit hochkonzentriertem Human-Fibrinogen, *Z. Kinderchir.,* 21, 1, 1981.

54. **Kuderna, H., Dinges, H. P., and Redl, H.,** Die Fibrinklebung in der Mikrochirurgie der peripheren Nerven, *H. Unfallheilkunde,* 148, 822, 1980.

55. **Kuderna, H.,** Die Fibrinklebung von Nervenanastomosen, in *Fibrinkleber in Orthopädie und Traumatologie,* Cotta, H. and Braun, A., Eds., Georg Thieme Verlag, Stuttgart, 1982, 254.

56. **Redl, H. and Schlag, G.,** Fibrinkleber—Methoden der Anwendung, in *Fibrinkleber in Orthopädie und Traumatologie,* Cotta, H. and Braun, A., Eds., Georg Thieme Verlag, Stuttgart, 1982, 286.

57. **Redl, H., Schlag, G., and Bancher, E.,** Grundlagen der Fibrinklebung, in *Fibrinkleber in der Orthopädie und Traumatologie,* Cotta, H. and Braun, A., Eds., Georg Thieme Verlag, Stuttgart, 1982, 18.

58. **Linscheer, W. G. and Fazio, T. L.,** Control of upper gastrointestinal hemorrhage by endoscopic spraying of clotting factors, *Gastorenterology,* 77, 642, 1979.

59. **Prindun, H. and Heindl, W.,** Fibrinspaytherapie beim Spontanpneumothorax, *Wr. Klin, Wrschr.,* 133(75), 40, 1983.

60. **Ferry, J. D. and Morrison, P. R.,** Fibrin Clots and Methods for Preparing the Same. U.S. Patent 2,533,004, 1950.

61. **Braun, A., Schumacher, G., Kratzat, R., Heine, W. D., and Pasch, B.,** Der Fibrin-Antibiotika-Verband im Tierexperiment zur lokalen Therapie des Staphylokokken infizierten Knochens, *H. Unfallheilkunde,* 148, 809, 1980.

62. **Bösch, P., Lintner, F., and Braun, F.,** Die autologe Spongiosatransplantation unter Anwendung des Fibrinklebsystems im Tierexperiment, *Wr. Kling. Woschr .,* 91, 628, 1979.

63. **Redl, H. Schlag, G., Stanek, G., Hirschl, A., and Sellich, T.,** In vitro properties of mixtures of fibrin seal and antibiotics, *Biomaterials,* 4, 29, 1983.

64. **Stanek, G., Bösch, P., Weber, P., and Hirschl, A.,** Experimentelle Untersuchungen über das pharmakokinetische Verhalten lokal applizierter Antibiotika, *Acta Med. Aust.,* 4(20), 19, 1980.

65. **Wahlig, H., Dingeldein, E., Braun, A., and Kratzat, R.,** Fibrinkleber und Antibiotika — Untersuchungen zur Freisetzungskinetik, in *Fibrinkleber in Orthopädie und Traumatologie,* Cotta, H. and Braun, A., Eds., Georg Thieme-Verlag, Stuttgart, 1982, 182.

66. **Richling, B.,** Homologous controller-viscosity fibrin for endovascular embolization. I. Experiment development of the medium, *Acta Neurochir.,* 62, 159, 1982.

67. **Dinges, H. P., Redl, H., and Schlag, G.,** Histologische Techniken und tierexperimentelle Modelle zur Untersuchung der Wechselwirkung zwischen Fibrinkleber und Gewebe, in *Fibrinkleber in Orthopädie und Traumatologie,* Cotta, H. and Braun, A., Eds., Georg Thieme Verlag, Stuttgart, 1982, 44.

68. **Heine, W. D., Braun, A., and Edinger, D.,** Gewebliche Abwehrreaktionen auf das Fibrinklebesystm, in *Fibrinkleber in Orthopädie und Traumatologie,* Cotta, H. and Braun, A., Eds., Georg Thieme Verlag, Stuttgart, 1982, 277.

69. **Marczell, A., Efferdinger, F., Spoule, J., and Stierer, M.,** Anwendungsbereiche des Fibrinklebers in der Abdominalchirurgie, *Acta Chir. Aust.,* 11, 137, 1979.

70. **Höllerl, G.,** Versorgung der verletzten Milz mittels Fibrinklebung, Infrarot-Kontakt-Koagulation und Laserkoagulation — Vergleichende tierexperimentelle Studie, *Acta Chir. Aust.,* Suppl. 37, 1, 1981.

71. **Scheele, J., Mühe, E., and Wopfner, F.,** Fibrinklebung — Eine neue Behandlungsmethode beim persistierenden und rezidivierenden Spontanpneumothorax, *Chirurg,* 49, 236, 1978.

72. **Hartel, W. and Laas, J.,** Zusätzliche Abdichtung von Pleura-Lunge-Läsionen mit Fibrinkleber, in *Neues über Fibrinogen, Fibrin und Fibrinkleber,* Blümel, G. and Haas, S., Eds., F. K. Schattauer Verlag, Stuttgart, 1981, 311.

73. **Akrami, R., Kalmar, P., Pokar, H. and Tilsner, V.,** Abdichtung von Kunststoffprothesen beim Ersatz der Aorta im thorakalen Bereich, *Thoraxchir. Vaskul. Chir.,* 26, 144, 1978.

74. **Haverich, A., Walterbusch, G., and Borst, H. G.,** The use of fibrin glue for sealing vascular prostheses of high porosity, *Thorac. Cardiovasc. Surg.,* 29, 251, 1981.

75. **Köveker, G., DeViviee, R., and Hellberg, K. D.,** Clinical experience with fibrin glue in cardiac surgery, *Thorac. Cardiovasc. Surg.,* 29, 251, 1981.

76. **Schuh, E., Braun, F., and Kovac, W.,** Knochenersatz bei schwerer progressiver Parodontopathie unter Anwendung des Fibrinklebesystems, *Öst. Zschr. Stomatol.,* 75, 411, 1978.

77. **Matras, H. and Jesch, W.,** Die Anwendung des Fibrinklebesystems zur Versorgung pathologischer Hohlräume im Kieferknochenbereich, *Dtsch. Z. Mund-Kiefer-Gesichtschir.,* 3, S43, 1979.

78. **Haas, S., Stemberger, A., Fritsche, H. M., Siegle, M., Tauber, R., and Blümel, G.,** Anwendung der Fibrinklebung bei verschiedenen hämorrhagischen Diathesen, in *Fibrinogen, Fibrin und Fibrinkleber,* Schimpf, K., Ed., F. K. Schattauer Verlag, Stuttgart, 1980, 279.

79. **Niekisch, R.,** Die Fibrinklebung — eine neue Methode zur Versorgung von Zahnextraktionswunden bei Patienten mit erhöhter Blutungsneigung, *Stomatol. DDR,* 30, 902, 1980.

80. **Siegle, M. and Brachmann, F.,** Die Behandlung blutungsgefährdeter Patienten mit Hilfe der Fibrinklebung, *Schweiz. Mschr. Zahnheilk.,* 90, 208, 1980.

81. **Nowotny, D. and Wurka, P.,** Anwendung von Fibrinkleber zur Blutstillung nach Zahnextraktionen bei Patienten mit angeborenen und erworbenen Gerinnungsstörungen, in *Fibrinogen, Fibrin und Fibrinkleber,* Schimpf, K., Ed., F. K. Schattauer Verlag, Stuttgart, 1980, 285.

82. **Kletter, G., Matras, H., and Dinges, H. P.,** Zur partiellen Klebung von Mikrogefäßanastomosen im intrakraniellen Bereich, *Wr. Klin. Woschr.,* 90, 415, 1978.

83. **Bösch, P., Braun, F., and Spängler, H. P.,** Die Technik der Fibrin-Spongiosaplastik, *Arch. Orthop. Unfall-Chir.,* 90, 63, 1977.

84. **Bösch, P., Nowotny, C., Schwägerl, W., Leber, H.,** Über die Wirkung des Fibrinklebesystems bei orthopädischen Operationen an Hämophilen und bei anderen Blutgerinnungsstörungen, in *Fibrinogen, Fibrin und Fibrinkleber,* Schimpf, K., Ed., F. K. Schattauer Verlag, Stuttgart, 1980, 275.

85. **Paar, O., Deigentesch, N., Prister, A., Riel, K. A., and Bernett, P.,** Fibrinklebung frischer Knorpelverletzungen am Talus, *Akt. Traumatol.,* 12, 19, 1982.

86. **Stübinger, B., Fritsche, H. M., Stemberger, A., Böttger, I., Prokscha, G. W., Pabst, H. W., and Blümel, G.,** Behandlung von Frakturen und knöchernen Defekten mit der ''Spongiosa-Fibrinkleber-Plombe'', in *Neues über Fibrinogen, Fibrin und Fibrinkleber,* Blümel, G., and Haas, S., Eds., F. K. Schattauer Verlag, Stuttgart, 1981, 357.

87. **Rauchenwald, K., Henning, K., and Urlesberger, H.,** Humanfibrinklebung bei Nephrotomie, *Helv. Chir. Acta,* 45, 283, 1978.

88. **Henning, K., Pflüger, H., Rauchenwald, K., and Urlesberger, H.,** Klininsche Erfahrungen mit Fibrinkonzentrat in der Nierenparenchymchirurgie, *Helv. Chir. Acta,* 47, 297, 1980.

89. **Rauchenwald, K., Urlesberger, H., Henning, K., Braun, F., Spängler, H. P., and Holle, J.,** Anwendung eines Fibrinklebers bei Operationen am Nierenparenchym, *Akt. Urol.,* 7, 209, 1976.

90. **Gasser, G., Mossig, H., Hopmeier, P., Lurf, M., and Fischer, M.,** Über die Verwendung eines Fibrinklebers bei Prostatektomie: Vorläufige Mitteilung, in *Fibrinogen, Fibrin und Fibrinkleber,* Schimpf, K., Ed., F. K. Schattauer Verlag, Stuttgart, 1980, 271.

91. **Frey, M., Holle, J., Mandl, H., and Freilinger, G.,** Die Vorteile der aufgeschobenen Spalthaut-Transplantation und die Erweiterung ihres Anwendungsbereiches durch die Verwendung des Fibrinklebers, *Acta Chir. Aust.,* 11, 97, 1979.

92. **Staindl, O.,** Die Gewebeklebung mit hochkonzentriertem, humanen Fibrinogen am Beispiel der freien, autologen Hauttransplantation, *Arch. Oto-Rhino-Laryngol.,* 217, 219, 1977.

93. **Genz, H. J. and Ludwig, H.,** Weiters Erfahrungen mit der Fibrinklebung bei schwangeren Frauen und vorzeitigem Blasensprung, in *Neues über Fibrinogen, Fibrin und Fibrinkleber,* Blümel, G. and Haas, S., Eds., F. K. Schattauer Verlag, Stuttgart, 1981, 341.

94. **Gastpar, H., Kastenbauer, E. R., and Behbehani, A. A.,** Erfahrungen mit einem humanen Fibrinkleber bei operativen Eingriffen im Kopf-Hals-Bereich, *Laryngol. Rhinol.,* 58, 389, 1979.

95. **Darf, W.,** Erfahrungen mit der Technik der Fibrinklebung in der Hals-Nasen-Ohren-Chirurgie, *Laryngol.Rhinol.,* 59, 99, 1980.

96. **Staindl, O.,** Beitrag zur chirurgischen Behandlung gutartiger Trachealstenosen, *Wr. Med. Woschr.,* 128, 175, 1978.

97. **Panis, R., Haid, T., and Scheele, J.,** Die Anwendung eines Fibrinklebers zur Rekonstruktion der hinteren Gehörgangswand, *Laryngol. Rhinol.,* 58, 400, 1979.

98. **Rendl, K. H., Staindl, O., Zelger, J., and Chmelizek-Feurstein, C.,** Die Hauttransplantation mit Fibrinkleber beim Ulcus cruris, *Akt. Dermatol.,* 6, 199, 1980.

99. **Köstering, H., Kasten, U., Artmann, U., Ruskowski, H., and Beyer, J. H.,** Behandlung von Hautnekrosen nach der Extravasation von Zytostatika mit Fibrinkleber, in *Neues über Fibrinogen, Fibrin und Fibrinkleber,* Blümel, G. and Haas, S., Eds., F. K. Schattauer Verlag, Stuttgart, 1981, 349.

100. **Scheele, J., Schricker, K. T., Goy, D., Lampe, I., and Panis, R.,** Hepatitisrisiko der Fibrinklebung in der Allgemeinchirurgie, *Med. Welt,* 32, 783, 1981.

101. **Panis, R. and Scheele, J.,** Hepatitisrisiko bei der Fibrinklebung in der HNO-Chirurgie, *Laryngol. Rhinol. Otol.,* 60, 367, 1981.

102. **Redl, H., Schlag, G. and Dinges, H. P.,** unpublished.

103. **Redl, H., Schlag, G., and Dinges, H. P.,** unpublished.

# INDEX